BASIC hydrology

Butterworths BASIC Series includes the following titles:

BASIC hydraulics
BASIC soil mechanics
BASIC thermodynamics and heat transfer
BASIC numerical mathematics
BASIC matrix methods
BASIC statistics
BASIC stress analysis

BASIC hydrology

James J Sharp ARCST, BSc, MSc, PhD, FICE, MCSCE, PEng, CEng
Professor of Engineering
Memorial University of Newfoundland

and

Peter G Sawden BEng
Principal, Battery Software Consulting Services Ltd

Butterworths
London Boston Durban Singapore Sydney Toronto Wellington

First published, 1984

© Butterworth & Co. (Publishers) Ltd, 1984

British Library Cataloguing in Publication Data

Sharp, James J.,
　　BASIC hydrology.
　　1. Hydraulic engineering—Data processing
　　2. Basic (Computer program language)
　　I. Title　　II. Sawden, Peter G.
　　627′.028′5424　TC157.8

　　ISBN 0-408-01363-X

Library of Congress Cataloging in Publication Data

Sharp, J. J. (James J.)
　　BASIC hydrology

　　Includes bibliographies and index.
　　1. Hydrology. I. Sawden, Peter G. II. Title.
GB661.2.S52　1984　　551.48　　83-14964
ISBN 0-408-01363-X

Typeset by Tunbridge Wells Typesetting Services Ltd
Printed and bound in Great Britain by The Thetford Press Ltd, Thetford, Norfolk

12/10/84

Preface

In common with other texts in the series, this book combines the application of BASIC programming with an engineering discipline, namely hydrology. Only a brief introduction is given to the use and application of BASIC (Chapter 1), but this is sufficient to permit a beginner to understand and to write BASIC programs. Many excellent, well written, books providing full details of the language are available and some of these are listed in the bibliography (p. 12). Following chapters deal primarily with hydrological topics. These are explained in some depth and worked examples developing programs for the solution of typical problems are provided. Once again, sufficient information is given to permit the student of hydrology to gain considerable understanding of the subject. Most of the topics considered are those which would be dealt with in an undergraduate introductory course. Further details and more exhaustive descriptions are provided in texts such as those listed in the bibliography.

Although the coverage of BASIC programming and hydrology is not as exhaustive as would be found in specialty texts, sufficient explanation of both is given to permit the text to be used by someone with no knowledge of either. The book should therefore be found useful by students in an undergraduate program or by practising engineers who are attempting to get to grips with modern computational procedures. The programs developed in the worked examples have been deliberately kept simple so that they may be easily understood and reproduced with the minimum of trouble. In many cases, however, the basic procedures used can be developed to provide much more sophisticated programs such as those which might be required for practical work.

Too often in the past, university students have received their primary exposure to computers through the use of large, complicated programs with vast amounts of data and a batch processing system with a turnround of as much as twelve hours. In such situations, the result was often disappointment, frustration, and a desire to avoid computers at all costs in the future. The advent of time sharing systems and in particular the development of microcomputers, has changed all this. Now it is possible to use a fairly powerful system and to have immediate contact with the computer. Many universities and consulting businesses have purchased personal microcomputers and students and practitioners alike are beginning to discover that computing can be fun. This text provides an introduction which can be used by those who have not yet reached that conclusion.

JJ Sharp, PG Sawden
St John's, Newfoundland, Canada
1983

Contents

Chapter 1

Introduction to BASIC

1.1 Introduction

BASIC is an acronym for Beginners All-purpose Symbolic Instruction Code. It was developed at Dartmouth College in the 1960s by Professors Kemeny and Kurtz for use on a time sharing system. The language is user oriented and makes use of instructions resembling basic algebraic formulae augmented by certain easily understood words such as LET, GO, TO, READ, PRINT, IF, THEN, etc. With the advent of the small, relatively cheap, personal computer, BASIC is now one of the most popular and most widely used of all computer languages. Other popular time sharing languages include ALGOL, COBOL, FORTRAN II, and FORTRAN IV. More specialised languages also exist. However, of all these, BASIC is the easiest to learn and use.

A BASIC program consists of a series of statements or lines, each beginning with a line number followed by a BASIC command. Except when instructed to do otherwise, the computer executes each line in order, beginning at the smallest line number and proceeding to the largest number at the end of the program. Statements do not have to be typed in order when the program is written. Figure 1.1

```
100  INPUT R
200  LET A = 3.142 ꞏ R ꞏ R
300  PRINT A,R
400  END
```
Figure 1.1

illustrates a simple BASIC program designed to calculate the area of a circle. The program consists of four statements and includes the BASIC keywords, INPUT, LET, PRINT, END. The first statement (100) allows the operator to specify the radius of the circle. The second statement (200) calculates the area A. The third statement prints the area and the radius, and the last statement identifies the end of the program. The key words used in this simple program, and other key words which permit alternative operations, will be described later in more detail.

1

Various steps must be followed in solving any program. It is important that the problem be clearly defined and that the user understand the problem completely. Following definition the problem must be described in mathematical terms and a formula or algebraic procedure, i.e. an algorithm, must be developed. When this has been done, a flow chart should be drawn to illustrate symbolically the logic of the solution. Then, following these three steps, the program itself is developed using the flowchart as a guide to the actual coding. To solve a specific problem the program must then be run on a computer. The first few runs should be used for checking purposes to ensure that the program did in fact do what was intended. This check is most important because it is easy to make errors in syntax or, in a complicated program, to have logic errors present. Finally, when the program is working properly, it is useful to develop some documentation describing it. The necessity for this documentation may not be immediately obvious but, if the program is stored for long periods between use it is very easy to forget the details required for satisfactory operation.

1.2   Flowcharts

The development of a flowchart assists the programmer to clarify the logic of the program. It specifies the operations which must be carried out by the computer and lists these in order. For example a flowchart to determine the area of a circle based on a specified diameter might take the form shown in Figure 1.2. If several calculations were to be performed using different diameters the

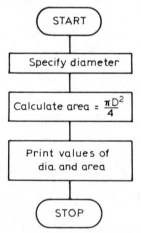

Figure 1.2   Flow chart for calculation of area

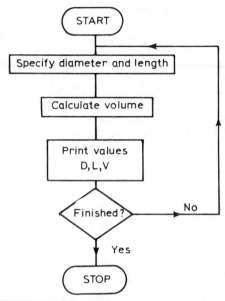

Figure 1.3   Flow chart for volume calculation

flowchart could be modified with a test to determine whether the program should be stopped or re-executed with a different diameter. Figure 1.3 shows such a flowchart developed to determine the volume of a cylinder of length L and diameter D. One simple way of determining when to stop calculations might be to specify zero diameter and to stop the program if the volume is equal to zero but to continue it for all non-zero cases. Such a program is shown in Figure 1.4. Statement 60 transfers control to line 100 when the volume is

```
20  INPUT D
30  INPUT L
40  LET V = 3.142 * D * D * L / 4.0
50  PRINT V,D,L
60  IF V = 0 THEN GOTO 100
70  GOTO 20
100 END
```
                                    Figure 1.4

zero. If the volume is not zero the program continues to line 70 which transfers control back to the beginning and requests the user to specify another value of D. To end the series of calculations the user would, at this time, specify a zero value for D or L, or both. Flowcharts may be drawn on paper of any size but it is important that the paper be large enough for clarity.

1.3   Variables, arrays and strings

Numerical quantities in BASIC are referred to as numbers or constants. They can be represented as integers (whole numbers with no decimal) or real constants which always include a decimal and may or may not have a fractional part. Very large or very small real constants may be expressed with an exponent. For example, the number 7521.3 may be expressed as 7.5213×10^3 and in BASIC would be written as 7.5213E3. Typically, constants may range in value from 10^{-38} to 10^{38} although the limits vary from one form of BASIC to another. The exponent must be an integer and may be positive or negative.

A simple variable is a name used to represent a number. It may consist of a letter or a letter followed by a digit and in some forms of BASIC two letters may be used. In Figure 1.4 the letters D, L and V are the variables. These can take different numerical values during execution of the program. More than two letters may be used for variable names but the computer will only recognise the first two. Thus, DOWN and DOTT would refer to one variable only.

A string is a sequence of letters and numbers and may vary in length from 15 characters to over 4000 characters depending on the particular version of BASIC. A string variable, used to represent a string, consists of a letter followed by a dollar sign. These are useful for including text in a program. For example, the test of whether a program is finished or whether calculations must be repeated, could be undertaken by printing a question 'ARE YOU FINISHED?' and requiring the user to type 'YES' or 'NO'. If this were applied to the problem shown in Figure 1.3, lines 60 and 70 of Figure 1.4 would be replaced by the four lines shown in Figure 1.5. Line 60 asks the user if he is finished. Line 70 requests the entry of a string variable, and line 80 tests to see if that string variable is 'YES', in which case the program is stopped.

```
60  PRINT "ARE YOU FINISHED ?  YES/NO"
70  INPUT B$
80  IF B$ = "YES" THEN GOTO 100
90  GOTO 20                          Figure 1.5
```

A list, or table, of values is referred to as an array. For example, the area of a circle may be required for 10 different values of diameter. These diameters could then be specified as D1, D2, . . . D10. The array in this case is one-dimensional and is simply a list of the diameters. Any particular item on the list can be specified by its number, and this can be represented in a BASIC program by an integer variable. For example, to undertake the calculations of circle

```
50  DIM R(10), A(10)
80  FOR I = 1 TO 10
100 INPUT R(I)
200 LET A(I) = 3.142 * R(I) * R(I)
300 PRINT A(I), R(I)
350 NEXT I
400 END
```
Figure 1.6

area using ten different radii, Figure 1.1 could be modified as shown in Figure 1.6. Lines 100 to 400 remain essentially the same but are written now as arrays. Line 80 sets up a counter and varies the integer I from 1 to 10 so that the calculation is repeated ten times. The loop for calculation is set up between lines 80 and 350. Line 50 is a dimension statement used to reserve the appropriate amount of space. Many formulations of BASIC permit arrays with up to ten items without a dimension statement. Two-dimensional arrays are used to handle tabulated data. Tables consist of horizontal rows and vertical columns. In specifying a two-dimensional array, the variable must be defined by two integers. For example, the variable $X(I, J)$ represents the value stored in row I and column J.

1.4 Input, output and format

Figures 1.1 to 1.6 demonstrate one method of entering data into the program. This is the INPUT statement which permits data to be entered while the program is being run. The most general form is that shown in Figure 1.1, line 100. The statement consists of the line number and the word INPUT, followed by the variable, R, which represents the number entered by the operator. When the statement is executed the computer types a question mark and waits until the user has typed in the value which is to be assigned to the variable R.

An alternative method of entering numeric values is to make use of the READ statement together with a DATA statement. The READ statement assigns a specific numeric value to a simple or array variable or assigns a string to a string variable. The values or string to be assigned are contained in a DATA statement. For example the statements

```
10  READ X,Y,Z
```

```
20  DATA 40,32.5,236.1
```

would cause the variable X to be assigned a numeric value of 40. The variable Y would be given a value 32.5 and the variable Z would be set equal to 236.1. DATA statements must be typed within the program but may be situated at any location. Notice that the order in which the data are specified is very important. Another form of READ statement is

```
30  READ A$,B$
```

where A$ and B$ are string variables. Simple, subscripted and string variables may be combined in the same READ statement. Thus the statement

```
40  READ X,Y,A$,B$,Z(1)
```

is quite valid.

The computer is made to output data by the use of the PRINT statement. This can be used to print the value of a variable or text depending on whether or not the part of the statement after PRINT is enclosed in inverted commas. For example, the statement

```
20  PRINT A,B
```

would cause the numeric values of variables A and B to be printed out separated by a space. The alternative version

```
20  PRINT "VALUES OF A AND B = "
```

would cause the computer to print out

```
VALUES OF A AND B =
```

These two types of PRINT statements may be combined in the following form

```
40  PRINT "VALUE OF A = "; A
```

If the variable A has previously been assigned a value of 6.32 the execution of the above statement causes the computer to print out

```
VALUE OF A = 6.32
```

Arithmetic expressions may also be included in the PRINT statement, for example,

```
40  PRINT "SQUARE OF A + B IS "; (A+B) * (A+B)
```

The PRINT statement used on its own without a following expression causes the computer to skip a line. In effect a blank line is printed and this is useful for separating parts of the output to enhance clarity.

REMARK statements enable the programmer to insert material into the program which is used for clarification but which does not form part of the program. This can be very useful in complicated programs and can assist anyone reading the program to understanding the operations which are being undertaken. For

```
10 REM CALCULATION OF CYLINDER VOLUME
35 REM   CALCULATION
45 REM   PRINT DATA
55 REM   TEST FOR COMPLETION           Figure 1.7
```

example, Figure 1.7 shows a series of REMARK statements which
could be added to Figure 1.4 in order to describe the various parts of
the program. As the computer ignores everything in a line following
the first three letters of REMARK, this makes the letters REM
interchangeable with REMARK.

1.5   Expressions, functions and subroutines

Mathematical expressions and BASIC expressions are quite similar.
They consist essentially of a combination of the operations of
addition, subtraction, multiplication, division, and exponentiation.
The operators used to accomplish these operations are shown in
Figure 1.8. The order in which operations are carried out, i.e. the

Operator	Operation	Example
+	Addition	A + B
—	Subtraction	A — B
*	Multiplication	A * B
/	Division	A / B
↑ or ∧	Exponentiation	A ↑ B

Figure 1.8 Mathematical operators

hierarchy of operations, is important to avoid confusion. Where
parentheses are used the operations contained within parenthesis are
performed first. Otherwise arithmetic operations are undertaken in
a fixed order of precedence. Exponentiation operations are
performed first. These are followed by multiplication and division
and lastly by addition and subtraction. Operations of equal
precedence are undertaken from left to right. The confusion which
might arise if the precedence of operation is not taken into account
can be illustrated by the solution of a simple quadratic equation, for
example,

$$AX^2 + BX + C = 0$$

One root of this equation is given by

$$X = \frac{-B + (B^2 - 4AC)^{1/2}}{2A}$$

In BASIC this expression must be written using parenthesis, i.e.

```
20 LET X = ( -B + SQR ( B ✻ B - 4.0 ✻ A ✻ C )) / ( 2.0 ✻ A )
```

The confusion which would exist if the parentheses were omitted is obvious.

The primary computational statement is the LET statement used in Figures 1.1 to 1.6. The right-hand side of the expression following the = sign may be any numerical value or any arithmetic expression. When the variable is a string variable, the right-hand side may of course be a string. The LET statement assigns the numeric value of the expression to the variable. In many forms of BASIC the LET may be omitted so that in Figure 1.1 line 200 could read

```
200 A = 3.142 ✻ R ✻ R
```

Figure 1.9 contains a list of library functions. These are standard functions built into BASIC and can be used directly in a LET statement. For example, the statement

```
20 LET X = SIN(Y)
```

causes the sine of the angle Y, where Y is in radians, to be computed and assigned to the variable X. Most statements are self explanatory but note that the INT function truncates the decimal part of a number and assigns a value lower than the real number specified. For example, 12.9 would be truncated to 12 and −4.1 would be assigned a value −5. The TAB function allows the programmer to specify exactly the location of material printed out using a PRINT statement. This allows fairly precise formatting of output data. The use of a comma in formatting was discussed earlier.

Function	Operation
ABS	Determines the absolute value
ATN	Determines the arctangent
COS	Determines the cosine (angle in radians)
COT	Determines the cotangent (″)
EXP	Raises e to a power
INT	Converts to integer form
LOG	Determines the natural logarithm
SGN	Determines the sign + 1, 0 or −1
SIN	Determines the sine (angle in radians)
SQR	Determines the square root
TAB	Formatting function
TAN	Determines the tangent

Examples
Let Y = Sin(X)   (sets Y equal to the sine of X)
Let Y = SQR(X)   (sets y equal to the square root of X)

Figure 1.9   Typical library functions

If a program requires the execution of a number of statements on several different occasions. it is convenient to define these statements as a subroutine, and make use of the routine as necessary throughout the main program. For example, if, in a program, the area of a circle was to be calculated at a number of different points, then the program illustrated in Figure 1.1 might be used as a subroutine. In the main program, the statement

```
10  GOSUB 100
```

would transfer control to the program written in Figure 1.1 whenever line 10 is executed. Statement 400 in Figure 1.1 would, however, be rewritten as

```
400  RETURN
```

The RETURN in line 400 transfers control from the subroutine back to the line in the main program immediately following line 10, i.e. the line in which the subroutine is called.

1.6   Control statements

As indicated earlier, a BASIC program is executed in the sequence determined by the line numbers unless otherwise specified. The simplest statement for altering the sequence of execution is the GO TO or GOTO statement. Examples of use are shown in Figures 1.4 and 1.5. Whenever a GOTO statement is encountered, the computer transfers control immediately to the line number given on the right-hand side of the statement. The computed GOTO statement causes control to be transferred to one of a group of statements, the particular one being chosen on the basis of the integer value of an expression. For example, the statement

```
30  ON A # B  GOTO 100,200,300
```

would transfer control to statement 100, 200 or 300 depending on the value of the integer part of $A \times B$. If $A \times B = 1$ control is transferred to line 100. If $A \times B = 2$ to line 200, and if $A \times B = 3$ to line 300. Some forms of BASIC use this statement to transfer control if an error is encountered. The appropriate statement would then be

```
10  ONERR  GOTO 500
```

This statement transfers control to line 500 if an error occurs within the program. This can be convenient at times when it is desired to keep the program running instead of an error signal being flagged.

The END statement is used to indicate the end of a program and to transfer control back to the user. Until the introduction of microcomputers using BASIC, it was compulsory to finish every

BASIC program with an END statement. However, some BASIC languages are now available in which the END statement is not required. The STOP statement may be used to terminate execution at any point in the program. It may appear at any point in a program and stops execution at that point.

The IF-THEN statement is often used to set a condition in a GOTO statement. Thus

```
20  IF X = 5 THEN GOTO 200
```

transfers control to line 200 when X equals 5. If X is not equal to 5 the program continues directly to the next line number. In many forms of BASIC the GOTO part of the statement may be omitted. In some, the THEN part may also be omitted, but in that case, the GOTO must be present. The above statement considers the case when X is exactly equal to 5. Other conditions may also be included in this statement and these are shown in Figure 1.10.

Operator	Operation
=	Equal to
< >	Not equal to
<	Less than
< =	Less than or equal to
>	Greater than
> =	Greater than or equal to

Figure 1.10   Conditional operators

The IF-THEN statement also permits the use of expressions other than GOTO. For example, the statement

```
200  IF X > 5 THEN Y = 0
```

sets Y equal to zero on occasions when X is greater than 5. If X is not greater than 5 the variable Y maintains the value it had prior to line 200.

The use of a FOR statement in setting up a loop was demonstrated in Figure 1.6. The loop is initiated by statement 80 which sets a counter, I equal to one. The following statements, until line 350 is reached, are performed with that value of I. At line 350 the value of I is increased by one and control is transferred back to line 80. The process is repeated until I has been incremented to ten as indicated in line 80. Unless otherwise specified the running variable I will always increase by one unit. However, this can be changed as necessary. The statement

```
80  FOR I = 1 TO 10 STEP 2
```

would increase the value of I in units of 2 so that the calculation would be performed 5 times instead of 10. In cases where the counter itself is to form part of the calculation, it is often convenient to use non-integer values in the FOR statement and this is permissible in some forms of BASIC.

1.7 Files

The use of data files is really beyond the scope of an introduction to BASIC, but data files are particularly important in the hydrological context because of the large amounts of data which must be stored and processed. Records of rainfall or river flow may extend over many years and it is important to store them so that they may be easily retrieved for use in a number of different situations. These data may be permanently stored in a number of different ways depending on the computer system in use, for example, punched tape, magnetic tape, floppy disks, etc. Files must be structured and organised consistently, so that data may be easily updated or modified if necessary and so that it can be read for use in computational programs.

Data files, sometimes called text files, may be sequential or random in nature. In a sequential file the individual items are arranged sequentially one after the other whereas in a random file the records are of a specified fixed length which permits fast and easy access to any part of the file. The size of the space set aside for the record is specified in advance, whereas with a sequential file, the size of the record is unspecified and depends entirely on the length of the record which is entered. Because there is no specified record length sequential files must be read sequentially starting at the beginning. With random access files organised into records of fixed length, it is possible to access any record immediately using either a key value or the position of the record in the file.

Programs which enter or retrieve data from a file are written in the same way as normal BASIC programs except that the READ and INPUT statements have reference to the file instead of to a terminal or DATA statement. When the file is held separately from the program it is necessary to specify the file name and location in order to provide access to it. Similarly, a PRINT or WRITE statement is used to enter data into the file, for example, to update a record of river flows. Again, the file name must be stated prior to use of the access statements.

Unfortunately, there is no standard system for operating files and commands used to access files are machine dependent. The Apple II personal computer, for example, executes disk-operating system

```
5   REM   ENTER 0 IF NO MORE DATA
10  INPUT "FLOW" ; F
20  INPUT "DATE" ; D
200 D$ = "" : REM  THERE IS A CTRL-D BETWEEN THE QUOTES
300 PRINT D$;"OPEN RIVER"
400 PRINT D$;"WRITE RIVER"
500 PRINT F
600 PRINT D
700 PRINT D$;"CLOSE RIVER"
800 IF F <> 0 THEN 10
900 END
```

Figure 1.11

commands from within a BASIC program by printing a string which consists of CTRL-D followed by the command. Files are opened and closed using the statements OPEN and CLOSE. Figure 1.11 shows how these commands could be used to generate a file called RIVER holding records of flow and date. The program is very simple and somewhat clumsy but demonstrates the methodology with reasonable clarity. Statements 10 and 20 permit the data on flow and date to be entered in the usual way. Statement 300 opens the file and statement 400 instructs the computer that any PRINT statements refer to the file and not to the monitor or printer. Statement 700 closes the file and statement 800 determines whether more data is to be entered or whether the exercise has been completed. The data can be retrieved by a similar program using INPUT or READ statements.

1.8   Bibliography

Bartee, T.C., *BASIC Computer Programming,* Harper and Row, New York, (1981).
Gottfried, B.S., *Programming with BASIC,* Schaum's Outline Series, McGraw Hill Book Co., New York, (1975).
Spencer, D.D., *A Guide to BASIC Programming,* Addison-Wesley Pub. Co., Reading Mass., (1970).

Chapter 2

Elements of hydrology

Hydrology deals with water from its arrival on the land surface until it returns to the atmosphere by evaporation or to the oceans by surface or subsurface flow. It is studied by geographers as a pure science and by engineers as an applied science. Pure scientists wish to understand the physical world and to discover what happens and why it happens. A large part of engineering, however, deals with the application of this knowledge for engineering purposes.

Much of hydrology is concerned with measurement, correlation and prediction. For example, it is important to measure the precipitation which may occur in a particular locality. Prediction of floods or of extreme rainfall events is of considerable importance and the correlation between precipitation and stream flow has significance if the stream flow patterns must be predicted from measured precipitation. Hydrologic measurements may be used to determine such things as minimum flows in rivers and the amount of storage required for water supply during droughts; maximum flow through reservoirs for spillway design; rainfall or runoff on urban areas for design of gutters and sewers; evaporation from storage lakes; and evapo-transpiration from crops to determine irrigation requirements.

Table 2.1 World distribution of water

Location	Approx. volume (km³)	Percentage of total
Oceans	1.3×10^9	97.18
Ice caps and glaciers	29×10^6	2.17
Available groundwater	4.2×10^6	0.31
Unavailable groundwater	4.2×10^6	0.31
Freshwater lakes	125 000	0.009
Saline lakes	104 000	0.007
Other	80 000	0.006
Approx. total	1.34×10^9	

Adapted from Nace, R. L., 'Water of the world', *Nat. Hist.* **73**, No. 1 (Jan. 1964).

An indication of the distribution of the world's estimated water resources is shown in Table 2.1. Obviously there is adequate water for all people in the world (approximately 5.5×10^{11} litres per person). The problem is that the largest part (ocean water) is too saline for use and much of the remainder is tied up in ice caps and glaciers. Only 0.31% of the total resource is available in groundwater, less than a half mile deep and less than one hundredth of one percent is available in fresh-water lakes. Lake water amounts to approximately 10×10^6 gallons per person on earth. This seems a very large amount but much of it is in areas which are remote from those who desperately need it. Together, the United States and Canada share about 20% of the world's lake water in the Great Lakes system. However, without renewal by rainfall or groundwater these would be fully utilised in about forty-one years if used by the United States as its sole source of water supply.

The movement of water from the ocean by evaporation and then back to the ocean by precipitation, surface runoff and groundwater flow is shown diagramatically in Figure 2.1. This is known as the hydrological cycle. When precipitation occurs, the first water to fall on the land surface may be intercepted by vegetation and buildings (interception storage). Some of the water will fall on the ground and create small puddles in the depressions (depression storage). If the rain stops at this point, water may never penetrate the ground and will simply evaporate from the plant surfaces and from the puddles. It is more likely, however, that with continuing precipitation some water will infiltrate and penetrate the ground. Part of this will be absorbed by the soil and part will be taken up by the roots of plants.

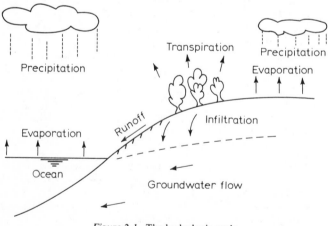

Figure 2.1 The hydrologic cycle

Some may, however, penetrate further and run into the zone of saturation below the water table. There it flows as groundwater towards streams and to the ocean. If the rain falls at a rate which is greater than the rate at which it can be absorbed into the ground (i.e. the intensity is greater than the infiltration capacity) surface runoff will occur. This is rarely seen on natural land surfaces but is very obvious whenever rain falls on inclined impermeable areas (e.g. roads, parking lots, etc.).

Engineering hydrology deals primarily with four parts of the hydrologic cycle. These are precipitation, surface runoff and stream flow, groundwater flow, and evaporation and transpiration.

2.1 Bibliography

Linsley, R.K., Kohler, M.A. and Paulus, J.L.H., *Hydrology for Engineers, 3rd edition,* McGraw Hill Book Co., New York, (1982).

Todd, D.K., *Groundwater Hydrology,* John Wiley and Sons, New York, (1980).

Viessman, W., Harbaugh, T.E. and Knapp, J.W., *Introduction to Hydrology,* Intext Ed. Pub., New York, (1972).

Wilson, E., *Engineering Hydrology 2nd edition,* MacMillan and Co. Ltd., London, (1974).

Chapter 3

Precipitation

ESSENTIAL THEORY

3.1 Introduction

Almost all precipitation originates from the sea. Evaporation lifts water vapour into the atmosphere where it circulates around the earth and, under the right conditions, falls back on to it. Depending on the temperature and location, precipitation may occur in a variety of forms: rain, fog, snow, sleet, hail, etc. Rain occurs when the temperature of an air mass drops to below the dew point temperature. At this point the air mass becomes saturated and, with further reduction in temperature, water vapour is precipitated out in liquid form.

Reasons for the cooling of an air mass vary, but in general there are three different types of mechanisms. These give rise to convective precipitation, orographic precipitation and cyclonic or frontal precipitation.

Convective precipitation occurs, as the name suggests, when a warm air mass rises. This may be due to unequal heating of the ground surface. As it rises, the air mass becomes cooler and precipitation may result if the temperature drops to the dewpoint. Orographic precipitation develops when air blowing in over the sea encounters high land and rises over it. This condition can often be seen in coastal regions where clouds sit around the peaks of a mountain range. Cyclonic precipitation results from the meeting of two air masses of different temperature. A 'front' refers to the boundaries of the air masses.

Figures 3.1 and 3.2 indicate the formation of warm and cold fronts. When a warm air mass overtakes a region of colder air, it will rise up on top of the cold air because of its lesser density (Figure 3.1). This situation is known as a warm front. An observer on the ground would notice the passage of a front by the following conditions. It would initially be relatively cold. Clouds would form overhead and precipitation might occur. Then as the front passes, the air would warm up significantly, the clouds would disappear and the weather

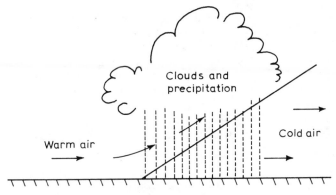

Figure 3.1   Warm front

would become reasonably bright and sunny. When a cold front passes, the conditions are quite different (Figure 3.2). Here, as cold air overtakes warm air it pushes in under it and again the warm air is forced to rise. However, an observer on the ground would start in relatively warm conditions and as the front passed, the temperature would drop, clouds would form and rain might begin.

In most temperate countries precipitation occurs mainly in the form of rain and is measured either by a standard rain gauge or by a recording gauge. Rain gauges consist essentially of cylindrical collectors with a chamfered edge. These collect the rain and direct it through an internal funnel into a container which may be removed for measurement purposes. The standard size and setting for the cylinder varies from country to country. Recording gauges provide a continuous record of the variation of rainfall with time. Generally

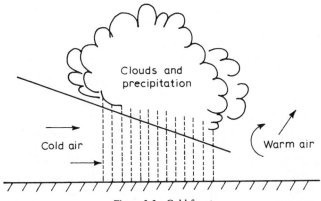

Figure 3.2   Cold front

they operate on the tipping bucket technique whereby a small container tips and discharges the rain whenever it becomes full. A record of each discharge is made on a clockwork-driven drum to give a time history of the rainfall. Such gauges are more prone to error than the standard gauge. It is important, therefore, that they should be used in conjunction with the standard gauge so that the totals may be correlated.

Correct interpretation of raw precipitation data is important in order to avoid errors in estimating the amounts of water available at any particular location.

3.2   Supplementing and checking records

In many cases, it is found that areas are inadequately recorded. For example, in a particular area there may be two rain gauges in similar surroundings; one, gauge A, recording and one, gauge B, requiring manual measurement and giving only the total. The record at B can be supplemented using the pattern at A. In Figure 3.3, the dotted line

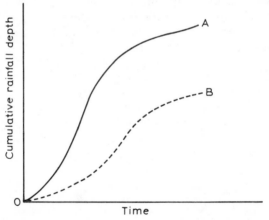

Figure 3.3   Supplementation of record at gauge B

from 0 to the maximum fall at B has been drawn to follow the pattern shown by gauge A and is probably better than a straight line. If records are missing because of instrument failure or lack of readings, it is possible to use a mathematical approach to compare the record at one gauge, A, with a number of nearby gauges, e.g. B, C and D. Several methods are available.

If the normal annual precipitation, N, at gauges B, C and D is

within 10% of that at gauge A, a straight average of the fall at the
three other gauges can be used to obtain the fall at A. If N at any of
the three index stations varies by more than 10% from that at gauge
A then a weighted average is used. Precipitation, P_A, at station A is
then given by

$$P_A = \frac{1}{3}\left(\frac{N_A}{N_B}P_B + \frac{N_A}{N_C}P_C + \frac{N_A}{N_D}P_D\right) \tag{3.1}$$

Another method uses four stations for an estimate. The area
around the station of interest is divided into four quadrants, north,
south, east and west. Records at the nearest station in each quadrant
are chosen for the calculation. A weighted average is again used but
in this case the weighting factor is given by the reciprocal of the

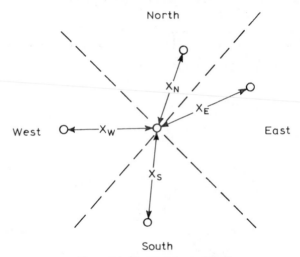

Figure 3.4   Four quadrant method

square of the distance. Thus for stations N, S, W and E, as indicated
in Figure 3.4, the precipitation at the central station is given by

$$P_X = \frac{1}{\Sigma 1/X^2}\left(\frac{P_N}{X_N^2} + \frac{P_S}{X_S^2} + \frac{P_E}{X_E^2} + \frac{P_W}{X_W^2}\right) \tag{3.2}$$

Multiple linear regression may also be applied to yield an equation
similar to (3.1) but with different constants.

   If a gauge is changed or moved to a new location, it is probable
that the precipitation catch relative to total precipitation will change.
For example, if the gauge is bumped it may move off the vertical and
thereafter be inclined at some small angle. If the rain falls vertically,

it would be expected that after the bump the gauge would catch less of the total rainfall than was previously the case. Inconsistencies in the record may also be caused by changes in methods of measuring or recording the data and can result from changes in personnel employed to collect data.

The double mass curve may be used to check the consistency of a particular record. Under normal circumstances, it would be expected that the cumulative fall at one, gauge A, would bear a fairly constant relationship to the cumulative fall at other nearby gauges. Thus a plot of the accumulated precipitation at A with the average (or summated) accumulated precipitation at a number of other nearby gauges should result in a straight line. Divergence from a straight line provides an indication of error at gauge A. The time at which the error occurred is indicated on the plot (see Figure 3.5) by the point at which the slope of the line changes. In Figure 3.5 this is shown to occur in 1960. Records after 1960 can be corrected by adjusting them according to the ratio of the slopes of the two lines.

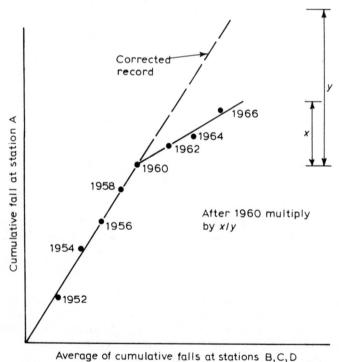

Figure 3.5   Use of double mass curve

Caution must be used in applying the double mass technique because the plotted points always fall about a mean line. Changes in slope should be identified only when these are significant.

3.3 Variation of depth with area

Many problems in hydrology require estimation of the average depth of fall over an area. A truly accurate estimation could be obtained only by covering the entire area in densely packed rain gauges so that the storm pattern could be defined precisely and in great detail. This of course would be impractical and quite uneconomic. Instead a number of gauges are scattered around the area in such a way as to give a reasonably accurate prediction. The minimum number of gauges required is determined by the size and physiography of the area and by typical storm conditions. For example, a mountainous area with steep slopes subjected to highly concentrated small storms would require more gauges than a relatively flat area over which the aerial distribution of rainfall was generally uniform. Table 3.1 provides an indication of typical densities of rain gauges for agricultural watersheds.

Table 3.1 Concentration of raingauges in agricultural watersheds

Size of drainage area (acres)	Minimum number of stations
0–30	1
30–100	2
100–200	3
200–500	1 per 100 acres
500–2500	1 per 250 acres
2500–5000	1 per sq. mile
over 5000	1 per each 3 sq. miles

After Holtan, H. N., Minshall, N. E. and Harrold, L. L., *Field Manual for Research in Agricultural Hydrology,* SWCD, ARS, Washington D.C., 214 pp. (1962).

There are basically three methods of estimating the depth over a catchment from records of point rainfall. The simplest, but least accurate, involves arithmetic averaging. In this method the average depth of fall over the entire catchment is assumed to be the average of the depths indicated on each rain gauge within the catchment boundary. This can provide a good estimate for flat catchments subjected to fairly uniform precipitation but can lead to errors if the rain gauges within the catchment show significantly different amounts. The Theissen Polygon takes account of the distribution of

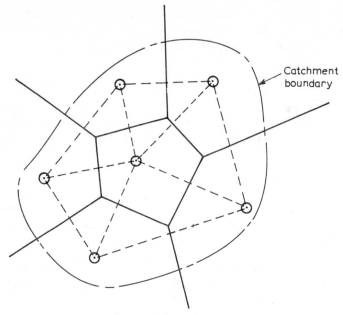

Figure 3.6   Theissen polygon

rainfall over the catchment and splits the catchment into a number of polygons each of which encloses one rain gauge. A weighted average of the gauge records is then calculated using the areas of the polygons as the weighting factors. Polygons are constructed as shown in Figure 3.6. The location of the rain gauges is plotted on a map and lines are drawn to join each station. These lines are then bisected with perpendiculars which are joined to form the polygons. This method is more accurate than simple averaging because it ties the rainfall at a particular gauge to an area surrounding that gauge. However, it assumes linear variation of rainfall between gauges and this may not always be the case. The third, and most accurate method, is the Isohyetal method. Isohyets are lines on a map joining points of equal rainfall. An isohyetal map is developed from the point recordings of rainfall in exactly the same manner in which a contour map is developed from spot heights. When isohyets have been drawn it may reasonably be assumed that the total volume of rain falling between any two isohyets is given by the product of the area between the isohyets and the average depth of fall. With experience this assumption may be slightly modified to take account of orographic effects and variations in the shape of particular storms. Curves showing the variation of depth with area over the catchment (depth-area curves) may be constructed as part of an isohyetal analysis. These show how the average depth of fall decreases as the area considered increases.

3.4   Variation of depth with time

The variation of rainfall with time may be shown in a variety of ways. If the depth of fall is obtained using a recording, tipping, bucket guage the record will show how the cumulative fall varies with time. If, on the other hand, a gauge is read on a daily basis, then the record obtained will show how much rain fell on each individual day. A plot of this data in the form shown in Figure 3.7 is known as a

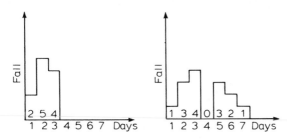

Hyetographs for 14-day period

Duration consec.days	Max. fall	Av. inten- sity
1	5	5
2	9	4.5
3	11	3.7
4	11	2.8
5	11	2.2
6	11	1.8
7	11	1.6

1st week

Duration consec.days	Max. fall	Av. inten- sity
1	4	4.0
2	7	3.5
3	8	2.7
4	10	2.5
5	12	2.4
6	13	2.2
7	14	2.0

2nd week

Figure 3.7   Depth-duration-intensity calculations

hyetograph or, alternatively, as a bar diagram. Data presented in this way can be adapted to show how the depth of fall varies with the duration of the storm, duration being defined as a continuous period of time during which rain falls. In constructing a depth-duration curve the record is searched to find the time period during which the maximum fall occurred. Then a search is made to find the maximum fall in two consecutive time periods, in three consecutive time periods and so on.

This procedure is shown in Figure 3.7 in which hyetographs are given for two consecutive weeks. Depth duration data for each week are also shown in the figure. Here the time period is one day. When constructing the table it is important to realise that duration refers to the number of consecutive days. Dividing the maximum depth of fall by the duration gives the average intensity. This is a measure of the rate at which precipitation occurs. From the tabulated data in Figure 3.7 it will be seen that the depth of fall increases with duration but that the average intensity decreases as the duration increases. This is what might be expected from visual observation of any storm.

Generally the rain starts off at a fairly low intensity, increases until the eye of the storm passes, and then decreases again. The average intensity over the duration of the whole storm is usually considerably less than the intensity at the height of the storm.

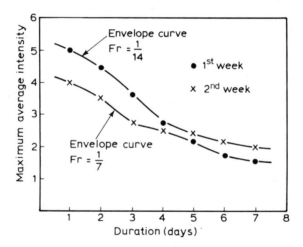

Figure 3.8   Intensity-duration-frequency

Figure 3.8 brings in the idea of frequency of occurrence and shows the variation of average intensity with duration using the data obtained in Figure 3.7. The top line in Figure 3.8 indicates an intensity which is equalled or exceeded in one day during the fourteen-day period. The lower line represents an intensity which is equalled or exceeded on two days in the fourteen-day period. Thus, the probability of obtaining an intensity given by the upper line is 1/14 whereas the lower line probability is $2/14 = 1/7$. This is known as the frequency of occurrence; defined as, the number of times within a certain period of time during which a specified occurrence (e.g. rainfall or flood) is equalled or exceeded. For example, if a certain flood was equalled or exceeded on two occasions in 100 years, the frequency would be 1/50. For predictive purposes it is important to realise that this does not mean that such a flood can be expected to occur only twice in any 100-year period and that if the flood is equalled or exceeded on the first year and second year of the period that the remaining 98 years will be flood free. The reciprocal of the frequency is known as the recurrence interval. This topic is referred to again in a separate chapter.

3.5   Analysis of trends

Trends may be real, caused by varying climatic conditions or may be apparent resulting from inconsistencies due to damage to a gauge or because a gauge has been shifted or replaced. The use of a double mass curve to check the consistency of a record has already been described.

Evidence of real trends may be obtained by a study of progressive long-term averages using five-year or three-year moving means. To develop these moving means the annual rainfall records at a particular station are examined. For a three-year moving mean t   ʼ

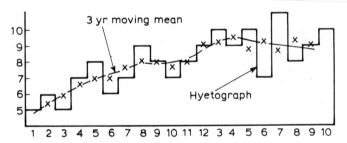

Figure 3.9   Three year moving mean

rainfall is averaged over successive three-year periods. The first mean would be obtained for years one, two and three and plotted at year two. The second mean would be obtained for years two, three and four and plotted at year three, etc. This is shown in Figure 3.9. Five-year moving means are prepared in the same way but with averages developed over five-year periods.

WORKED EXAMPLES

Example 3.1 DOUBLEMASS: consistency of records

The annual rainfall values for five stations all in the same catchment area are given below. Develop a program to facilitate use of the double mass technique in order to check the consistency of each of the five stations.

Year	1	2	3	4	5
1973	43.54	40.10	44.21	39.17	39.91
1974	48.80	47.54	48.41	43.34	45.15
1975	47.57	46.77	47.50	42.28	42.74
1976	43.15	43.26	43.86	35.02	33.12
1977	45.03	44.91	50.95	37.86	48.91
1978	45.99	47.06	43.10	37.36	37.15
1979	40.41	40.16	38.94	35.71	40.77
1980	63.77	61.75	60.57	52.23	54.07

The mass curve technique technique involves plotting the fall at one station against the sum, or the average, of the falls at a number of nearby stations. For example, to check the consistency of station 1, the fall at station 1 would be plotted against the average of the falls at stations 2, 3, 4 and 5 in each of the years from 1973 to 1980. Station 2 would be checked in the same way by plotting the fall at 2 against the average of the falls at 1, 3, 4 and 5, etc. The program will be developed to handle the computational aspects of this problem, the output being tabulated data which are suitable for plotting. In order to generalise the program it should be capable of handling any number of stations.

```
10 CLS
20 PRINT "DOUBLE MASS CURVE ANALYSIS"
30 PRINT "--------------------------"
40 PRINT
50 PRINT "ENTER NUMBER OF STATIONS TO BE USED "
60 INPUT S
70 PRINT
80 PRINT "ENTER NUMBER OF OBSERVATIONS AT EACH STATION"
90 INPUT N
100 DIM R(S,N)
110 FOR I = 1 TO S
120 FOR J = 1 TO N
130 PRINT "FOR STATION NUMBER "; I ;" - OBSERVATION "; J ;" IS :"
140 INPUT R(I,J)
150 NEXT J
160 NEXT I
170 REM START OF CALCULATIONS
180 PRINT "ENTER STATION NUMBER YOU WISH CALCULATED "
190 INPUT T
200 PRINT "STATION NO. "; T ;"    SUM OF OTHER STATIONS"
220 PRINT "--------------    ---------------------"
240 A = 0
250 B = 0
```

```
260 FOR L = 1 TO N
270 FOR J = 1 TO S
280 IF J = T THEN GOTO 320
290 A = A + R(J,L)
300 X = A / (S-1)
310 GOTO 330
320 B = B + R(J,L)
330 NEXT J
340 PRINT TAB(5) ; B ; TAB(28) ; X
360 NEXT L
370 PRINT
390 PRINT
410 PRINT "DO YOU WISH TO CALCULATE FOR ANOTHER STATION WITH THIS DATA BASE ( Y/N ) ? "
420 INPUT A$
430 IF A$ = "Y" THEN GOTO 180
440 END
```

DOUBLE MASS CURVE ANALYSIS

ENTER NUMBER OF STATIONS TO BE USED
? 5

ENTER NUMBER OF OBSERVATIONS AT EACH STATION
? 8

FOR STATION NUMBER 1 - OBSERVATION 1 IS :
? 43.54
FOR STATION NUMBER 1 - OBSERVATION 2 IS :
? 48.8
FOR STATION NUMBER 1 - OBSERVATION 3 IS :
? 47.57
FOR STATION NUMBER 1 - OBSERVATION 4 IS :
? 43.15
FOR STATION NUMBER 1 - OBSERVATION 5 IS :
? 45.03
FOR STATION NUMBER 1 - OBSERVATION 6 IS :
? 45.99
FOR STATION NUMBER 1 - OBSERVATION 7 IS :
? 40.41
FOR STATION NUMBER 1 - OBSERVATION 8 IS :
? 63.77
FOR STATION NUMBER 2 - OBSERVATION 1 IS :
? 40.1
FOR STATION NUMBER 2 - OBSERVATION 2 IS :
? 47.54
FOR STATION NUMBER 2 - OBSERVATION 3 IS :
? 46.77
FOR STATION NUMBER 2 - OBSERVATION 4 IS :
? 43.26
FOR STATION NUMBER 2 - OBSERVATION 5 IS :
? 44.91
FOR STATION NUMBER 2 - OBSERVATION 6 IS :
? 47.06
FOR STATION NUMBER 2 - OBSERVATION 7 IS :
? 40.16
FOR STATION NUMBER 2 - OBSERVATION 8 IS :
? 61.75

FOR STATION NUMBER 3 - OBSERVATION 1 IS :
? 44.21
FOR STATION NUMBER 3 - OBSERVATION 2 IS :
? 48.41
FOR STATION NUMBER 3 - OBSERVATION 3 IS :
? 47.5
FOR STATION NUMBER 3 - OBSERVATION 4 IS :
? 43.86
FOR STATION NUMBER 3 - OBSERVATION 5 IS :
? 50.96
FOR STATION NUMBER 3 - OBSERVATION 6 IS :
? 43.1
FOR STATION NUMBER 3 - OBSERVATION 7 IS :
? 38.94
FOR STATION NUMBER 3 - OBSERVATION 8 IS :
? 60.57
FOR STATION NUMBER 4 - OBSERVATION 1 IS :
? 39.17
FOR STATION NUMBER 4 - OBSERVATION 2 IS :
? 43.34
FOR STATION NUMBER 4 - OBSERVATION 3 IS :
? 42.28
FOR STATION NUMBER 4 - OBSERVATION 4 IS :
? 35.02
FOR STATION NUMBER 4 - OBSERVATION 5 IS :
? 37.86
FOR STATION NUMBER 4 - OBSERVATION 6 IS :
? 37.36
FOR STATION NUMBER 4 - OBSERVATION 7 IS :
? 35.71
FOR STATION NUMBER 4 - OBSERVATION 8 IS :
? 52.23

```
FOR STATION NUMBER  5  - OBSERVATION  1  IS :
? 39.91
FOR STATION NUMBER  5  - OBSERVATION  2  IS :
? 45.15
FOR STATION NUMBER  5  - OBSERVATION  3  IS :
? 42.74
FOR STATION NUMBER  5  - OBSERVATION  4  IS :
? 33.12
FOR STATION NUMBER  5  - OBSERVATION  5  IS :
? 48.91
FOR STATION NUMBER  5  - OBSERVATION  6  IS :
? 37.15
FOR STATION NUMBER  5  - OBSERVATION  7  IS :
? 40.77
FOR STATION NUMBER  5  - OBSERVATION  8  IS :
? 54.07
ENTER STATION NUMBER YOU WISH CALCULATED
? 1
STATION NO.  1    SUM OF OTHER STATIONS
---------------   ---------------------
     43.54            40.8475
     92.33999         86.95749
    139.91           131.78
    183.06           170.595
    228.09           216.255
    274.08           257.4225
    314.49           296.3175
    378.26           353.4725
DO YOU WISH TO CALCULATE FOR ANOTHER STATION WITH THIS DATA BASE ( Y/N ) ?
? Y
ENTER STATION NUMBER YOU WISH CALCULATED
? 2
STATION NO.  2    SUM OF OTHER STATIONS
---------------   ---------------------
     40.1             41.7075
     87.64            88.1325
    134.41           133.155
    177.67           171.9425
    222.58           217.6325
    269.64           258.5325
    309.8            297.49
    371.55           355.15
DO YOU WISH TO CALCULATE FOR ANOTHER STATION WITH THIS DATA BASE ( Y/N ) ?
? Y
ENTER STATION NUMBER YOU WISH CALCULATED
? 3
STATION NO.  3    SUM OF OTHER STATIONS
---------------   ---------------------
     44.21            40.68
     92.61999         86.8875
    140.12           131.7275
    183.98           170.365
    234.94           214.5425
    278.04           256.4325
    316.98           295.695
    377.55           353.65
DO YOU WISH TO CALCULATE FOR ANOTHER STATION WITH THIS DATA BASE ( Y/N ) ?
? Y
```

```
ENTER STATION NUMBER YOU WISH CALCULATED
? 4
STATION NO.  4    SUM OF OTHER STATIONS
---------------   ---------------------
    39.17               42.0175
    82.51               89.4925
   124.79              135.6375
   159.81              176.485
   197.67              223.9375
   235.03              267.2625
   270.74              307.3325
   322.97              367.3725
DO YOU WISH TO CALCULATE FOR ANOTHER STATION WITH THIS DATA BASE ( Y/N ) ?
? Y


ENTER STATION NUMBER YOU WISH CALCULATED
? 5
STATION NO.  5    SUM OF OTHER STATIONS
---------------   ---------------------
    39.91               41.8325
    85.06               88.833
   127.8               134.885
   160.92              176.2075
   209.83              220.8975
   246.98              264.275
   287.75              303.08
   341.82              362.66
DO YOU WISH TO CALCULATE FOR ANOTHER STATION WITH THIS DATA BASE ( Y/N ) ?
? N
```

Program notes

(1) Lines 0-40 — print titles.
(2) Lines 50-160 — operator inputs data (number of stations, number of observations at each station, all observations at station 1, then station 2, station 3, etc. until complete).
(3) Lines 180-230 — print titles and request operator to choose station to be checked.
(4) Lines 240-400 — calculate average of falls at all stations other than that chosen and print average against fall at chosen station.
(5) Lines 410-440 — permit calculation using other station choice.
Note: output data must be plotted to check for possible errors.

Example 3.2 FOURQUAD: estimation of missing rainfall

The record at rain gauge X is incomplete for a certain period of time but four rain gauges located as shown in Figure 3.4 do have complete records for that period of time. The falls at these stations and their distances from gauge X are given below. Determine the fall at station X.

Gauge	Fall (mm)	Distance from X (m)
N	34.5	1547
S	25.3	1986
W	29.1	1052
E	39.6	1126

The missing record may be estimated using the four quadrant method described earlier. Equation (3.2) applies.

```
5 DIM P(5),L(5),S(5),D$(5)
10 CLS
20 PRINT "WEIGHTED AVERAGE - FOUR QUADRANT METHOD"
30 PRINT
40 PRINT "ENTER PRECIPITATION AT STATION 'N','E','S','W' AND DISTANCE TO STATION 'X' WHEN PROMPTED"
50 D$(1) = "NORTH QUADRANT ',' DISTANCE"
60 D$(2) = "EAST QUADRANT  ',' DISTANCE"
70 D$(3) = "SOUTH QUADRANT ',' DISTANCE"
80 D$(4) = "WEST QUADRANT  ',' DISTANCE"
90 FOR I = 1 TO 4
100 PRINT
110 PRINT "PRECIP. AT STATION IN ";D$(I);" TO STATION 'X' ( 25 cm , 2000 m )"
120 INPUT P(I),L(I)
130 S(I) = L(I) ^ 2
140 NEXT I
150 T = 0
160 U = 0
170 FOR I = 1 TO 4
180 T = T + ( P(I)/S(I) )
190 U = U + ( 1/S(I) )
200 NEXT I
210 V = T/U
220 PRINT
230 PRINT "THE PRECIPITATION AT STATION 'X' USING WEIGHTED AVERAGE - 4 QUADRANT METHOD IS "; V ;" UNITS. "
240 PRINT
250 PRINT "DO YOU WISH TO USE THIS METHOD AGAIN ( Y/N ) ?"
260 INPUT A$
270 IF A$ = "Y" THEN GOTO 10
280 END


RUN

WEIGHTED AVERAGE - FOUR QUADRANT METHOD

ENTER PRECIPITATION AT STATION 'N','E','S','W' AND DISTANCE TO STATION 'X' WHEN
PROMPTED

PRECIP. AT STATION IN NORTH QUADRANT ',' DISTANCE
 TO STATION 'X' ( 25 cm , 2000 m )
? 34.5,1547

PRECIP. AT STATION IN EAST QUADRANT  ',' DISTANCE
 TO STATION 'X' ( 25 cm , 2000 m )
? 39.6,1126

PRECIP. AT STATION IN SOUTH QUADRANT ',' DISTANCE
 TO STATION 'X' ( 25 cm , 2000 m )
? 25.3,1986
```

```
PRECIP. AT STATION IN WEST QUADRANT  ',' DISTANCE
TO STATION 'X' ( 25 cm , 2000 m )
? 29.1,1052
```

```
THE PRECIPITATION AT STATION 'X' USING WEIGHTED AVERAGE - 4 QUADRANT METHOD IS
33.15066  UNITS.
```

```
DO YOU WISH TO USE THIS METHOD AGAIN ( Y/N ) ?
? N
```

Program notes

(1) Lines 0-5 — set dimensions for arrays.

(2) Lines 10-30 — print title.

(3) Lines 40-120 — request user to input fall and distance for stations in North, East, South and West quadrants.

(4) Lines 130-230 — calculate fall at central station using Equation (3.2) and print result.

(5) Lines 240-280 — permit recalculation using other data.

Example 3.3: INTENDUR: depth-intensity-duration analysis

The record of the daily fall at a rain gauge is shown below. Develop a program which can be used to plot depth duration and intensity duration curves.

Day	1	2	3	4	5	6	7	8	9	10	11	12
Fall (mm)	0	3	5	8	7	9	10	8	7	7	4	2

The program will be generalised to handle any number of observations. A search should be made for the maximum fall in any one day, in any two consecutive days, in any three consecutive days and so on. Data should be printed out showing maximum fall against duration and maximum average intensity against duration.

```
10 CLS
20 PRINT "INTENSITY - DURATION ANALYSIS"
30 PRINT "-----------------------------"
40 PRINT : PRINT : PRINT
50 PRINT "ENTER NUMBER OF OBSERVATIONS"
60 INPUT N
70 DIM R(N),S(N),T(N,N)
80 PRINT "ENTER UNITS OF PRECIPITATION (i.e. INCHES, mm , FEET)
90 INPUT B$
100 PRINT "ENTER TIME UNITS (i.e. MINUTES,HOURS,DAYS)"
110 INPUT M$
120 PRINT "ENTER TIME INTERVAL FOR OBSERVATIONS"
130 INPUT M
150 FOR I = 1 TO N
160 PRINT "ENTER OBSERVATION "; I ;" = ";
170 INPUT R(I)
180 NEXT I
```

```
190 C = 0
200 FOR J = 1 TO N
210 FOR L = 1 TO N - (J-1)
220 FOR I = 1 TO J
230 C = C + R(I + (L-1) )
240 NEXT I
250 T(J,L) = C
260 C = 0
270 NEXT L
280 NEXT J
290 C = 0
300 FOR J = 1 TO N
310 FOR L = 1 TO N - (J-1)
320 IF C < T(J,L) THEN C = T(J,L)
330 NEXT L
340 T(J,1) = C
350 NEXT J
360 PRINT : PRINT : PRINT
365 A = LEN(M$)
370 A$ = LEFT$(M$,A-1)
380 PRINT "DURATION (";M$;")";TAB(22);"TOTAL FALL (";B$;")";TAB(44);"MAX. INTENSITY (";B$;" per ";A$;")"
390 PRINT "--------------------------------------------------------------------------------"
400 FOR I = 1 TO N
410 PRINT TAB(5);M#I;TAB(27);T(I,1);TAB(54);T(I,1)/(M#I)
420 NEXT I
430 PRINT : PRINT : PRINT
440 END
```

RUN

INTENSITY - DURATION ANALYSIS

ENTER NUMBER OF OBSERVATIONS
? 12
ENTER UNITS OF PRECIPITATION (i.e. INCHES, mm , FEET)
? mm
ENTER TIME UNITS (i.e. MINUTES,HOURS,DAYS)
? Days
ENTER TIME INTERVAL FOR OBSERVATIONS
? 1
ENTER OBSERVATION 1   = ? 0
ENTER OBSERVATION 2   = ? 3
ENTER OBSERVATION 3   = ? 5
ENTER OBSERVATION 4   = ? 8
ENTER OBSERVATION 5   = ? 7
ENTER OBSERVATION 6   = ? 9
ENTER OBSERVATION 7   = ? 10
ENTER OBSERVATION 8   = ? 8
ENTER OBSERVATION 9   = ? 7
ENTER OBSERVATION 10 = ? 7
ENTER OBSERVATION 11 = ? 4
ENTER OBSERVATION 12 = ? 2

DURATION (Days)	TOTAL FALL (mm)	MAX. INTENSITY (mm per Day)
1	10	10
2	19	9.5
3	27	9
4	34	8.5
5	42	8.399999
6	49	8.166667
7	56	8
8	61	7.625
9	65	7.222223
10	68	6.8
11	70	6.363637
12	70	5.833334

Program notes

(1) Lines 0-40 — clear screen and print title.
(2) Lines 50-180 — input data; number of observations, units and values.
(3) Lines 190-350 — calculate maximum fall over full range of different durations.
(4) Lines 360-440 — print data tables and stop run.

Example 3.4: DEPAREA: depth-area analysis

Following a storm on a particular catchment, an isohyetal map is drawn. The total area enclosed by the isohyets is given below. Develop a program to calculate the variation of depth with area over the catchment.

Isohyet (mm)	100	75	50	25	<25
Total area enclosed (sq km)	32	224	500	1005	1517

The volume of rain falling on a particular area must be evaluated using the average depth between isohyets, coupled with the area between the same isohyets. In the area enclosed by the 100 mm isohyet it will be assumed that the average depth is 110 mm. For the area outside the 25 mm isohyet it will be assumed that the average depth is 20 mm.

```
10 CLS
20 REM AREAL AVERAGING OF PRECIPITATION - ISOHYETAL METHOD
30 PRINT "ISOHYETAL METHOD"
40 PRINT "----------------"
50 PRINT
60 PRINT "ENTER NUMBER OF ISOHYETAL AREAS "
70 INPUT N
80 PRINT
90 DIM L$(N),M(N+1),R(N),E(N),P(N),Q(N),D(N)
100 PRINT "ENTER INFORMATION AS PROMPTED:"
110 PRINT
120 PRINT
130 FOR I = 1 TO N
140 PRINT
150 PRINT
160 PRINT "FOR ISOHYETAL AREA "; I
170 PRINT
180 PRINT "ENTER ISOHYET (i.e. '100' or '< 10') "
190 INPUT L$(I)
200 PRINT "ENTER AREA ENCLOSED ( sq. km )"
210 INPUT M(I)
220 PRINT "ENTER AVERAGE PRECIPITATION ( mm )"
230 INPUT R(I)
```

```
240 NEXT I
250 PRINT
260 PRINT
270 PRINT
280 A = 0
290 B = 0
300 M(0) = 0
310 FOR I = 1 TO N
320 E(I) = M(I) - M(I-1)
330 P(I) = E(I) * R(I)
340 B = B + P(I)
350 Q(I) = B
360 D(I) = Q(I) / M(I)
370 NEXT I
380 PRINT "ISOHYET  AREA  NET AREA  AVG. PREC.  PREC. VOL.  TOTAL PREC. VOL.  AVG. DEPTH"
390 PRINT "         (sq km) (sq km)    (mm)       (cu m)        (cu m)          (mm) "
400 PRINT "-------  ----  -------  ----------  ----------  ----------------  ----------"
410 FOR I = 1 TO N
420 PRINT TAB(3);L$(I);TAB(10);M(I);TAB(18);E(I);TAB(28);R(I);TAB(40);P(I);TAB(54);Q(I);TAB(68);D(I)
430 NEXT I
440 PRINT
450 PRINT
460 PRINT
470 END
```

RUN

ISOHYETAL METHOD

ENTER NUMBER OF ISOHYETAL AREAS
? 5

ENTER INFORMATION AS PROMPTED:

FOR ISOHYETAL AREA 1

ENTER ISOHYET (i.e. '100' or '< 10')
? 100
ENTER AREA ENCLOSED (sq. km)
? 32
ENTER AVERAGE PRECIPITATION (mm)
? 110

FOR ISOHYETAL AREA 2

ENTER ISOHYET (i.e. '100' or '< 10')
? 75
ENTER AREA ENCLOSED (sq. km)
? 224.
ENTER AVERAGE PRECIPITATION (mm)
? 87.5

FOR ISOHYETAL AREA 3

ENTER ISOHYET (i.e. '100' or '< 10')
? 50
ENTER AREA ENCLOSED (sq. km)
? 500
ENTER AVERAGE PRECIPITATION (mm)
? 62.5

FOR ISOHYETAL AREA 4

ENTER ISOHYET (i.e. '100' or '< 10')
? 25
ENTER AREA ENCLOSED (sq. km)
? 1005
ENTER AVERAGE PRECIPITATION (mm)
? 37.5

FOR ISOHYETAL AREA 5

ENTER ISOHYET (i.e. '100' or '< 10')
? < 25
ENTER AREA ENCLOSED (sq. km)
? 1517
ENTER AVERAGE PRECIPITATION (mm)
? 20

ISOHYET	AREA (sq km)	NET AREA (sq km)	AVG. PREC. (mm)	PREC. VOL. (cu m)	TOTAL PREC. VOL. (cu m)	AVG. DEPTH (mm)
100	32	32	110	3520	3520	110
75	224	192	87.5	16800	20320	90.71429
50	500	276	62.5	17250	37570	75.14
25	1005	505	37.5	18937.5	56507.5	56.22637
(25	1517	512	20	10240	66747.5	43.99967

Program notes

(1) Lines 0-40 — print title.
(2) Lines 50-260 — enter data: number of areas considered, area and average precipitation between isohyets. (Line 90 sets array storage space.)
(3) Lines 270-370 — calculate net area, precipitation volume, total volume and average depth.
(4) Lines 380-400 — print table titles.
(5) Lines 410-430 — print calculated data.
(6) Lines 440-470 — stop execution of program.

Example 3.5: MOVMEAN: moving mean analysis

Annual rainfall data over ten years at a particular station are given below. Calculate the five-year and three-year moving means.

Year	1	2	3	4	5	6	7	8	9	10
Fall (mm)	922	711	731	786	620	600	889	459	716	652

Averages are taken over five-year and three-year periods beginning at year one and moving progressively through the record one year at a time.

```
10 CLS
20 REM MOVING MEAN
30 PRINT "MOVING MEAN PROGRAM"
40 PRINT "-------------------"
50 PRINT
60 PRINT "NUMBER OF OBSERVATIONS TO BE ENTERED"
70 INPUT N
80 DIM R(N),T(N),K(N)
90 FOR I = 1 TO N
100 PRINT "ENTER OBSERVATION "; I
110 INPUT R(I)
120 NEXT I
130 PRINT "ENTER NUMBER OF YEARS FOR MEAN  ( i.e. 3 or 5 )"
140 INPUT P
150 K = 1
160 FOR L = 1 TO ( N - (P-1) )
```

```
170 FOR I = 1 TO P
180 A = A + R( I + (L-1) )
190 NEXT I
200 T(K) = A/P
210 K = K + 1
220 A = 0
230 NEXT L
240 PRINT "THE "; P ;" YEAR MOVING MEAN IS :"
250 FOR I = 1 TO ( N - (P-1) )
260 K(I) = (( P/2) + .5 ) + ( I-1 )
270 PRINT "AT TIME";K(I); TAB(18) " EQUALS "; T(I)
280 NEXT I
290 PRINT
300 PRINT
310 PRINT
320 PRINT "DO YOU WISH TO DO ANOTHER PERIOD MOVING MEAN WITH SAME DATA ( Y/N ) ? "
330 INPUT A$
340 IF A$ = "Y" THEN GOTO 130
350 END
```

```
RUN


MOVING MEAN PROGRAM
-------------------

NUMBER OF OBSERVATIONS TO BE ENTERED
? 10
ENTER OBSERVATION  1            ENTER OBSERVATION  6
? 922                          ? 600
ENTER OBSERVATION  2           ENTER OBSERVATION  7
? 711                          ? 889
ENTER OBSERVATION  3           ENTER OBSERVATION  8
? 731                          ? 459
ENTER OBSERVATION  4           ENTER OBSERVATION  9
? 786                          ? 716
ENTER OBSERVATION  5           ENTER OBSERVATION  10
? 620                          ? 652



ENTER NUMBER OF YEARS FOR MEAN   ( i.e. 3 or 5 )
? 3
THE  3  YEAR MOVING MEAN IS :
AT TIME 2          EQUALS  788
AT TIME 3          EQUALS  742.6667
AT TIME 4          EQUALS  712.3333
AT TIME 5          EQUALS  668.6667
AT TIME 6          EQUALS  703
AT TIME 7          EQUALS  649.3333
AT TIME 8          EQUALS  688
AT TIME 9          EQUALS  609

DO YOU WISH TO DO ANOTHER PERIOD MOVING MEAN WITH SAME DATA ( Y/N ) ?
? Y
```

```
ENTER NUMBER OF YEARS FOR MEAN  ( i.e. 3 or 5 )
? 5
THE  5  YEAR MOVING MEAN IS :
AT TIME 3        EQUALS  754
AT TIME 4        EQUALS  689.6
AT TIME 5        EQUALS  725.2
AT TIME 6        EQUALS  670.8
AT TIME 7        EQUALS  656.8
AT TIME 8        EQUALS  663.2
DO YOU WISH TO DO ANOTHER PERIOD MOVING MEAN WITH SAME DATA ( Y/N ) ?
? N
```

Program notes

(1) Lines 0-50 — print title.

(2) Lines 60-120 — enter number of observations, set array storage space (line 80) and enter values.

(3) Lines 130-140 — set number of years for mean (3 or 5 in this case).

(4) Lines 150-230 — calculate mean.

(5) Lines 240-280 — print mean and central point for plotting.

(6) Lines 290-350 — permit recalculation with additional data.

PROBLEMS

(3.1) The data below provide details of individual storm precipitation at four gauges A, B, C and D together with normal annual precipitation. Develop a program, based on Equation (3.1) to estimate the missing data.

Gauge	A	B	C	D
Precipitation (mm)	—	50	58	35
Precipitation	25	—	20	21
Precipitation	35	25	29	—
Normal annual (mm)	1200	1000	700	810

(3.2) The data below give amounts of storm precipitation at five

Observed precipitation (mm)	Area (km)^2
16.5	18
23.9	305
49.3	260
68.0	310
38.5	53
74.9	234
126.3	204
103.2	195

individual gauges together with the areas of Theissen polygons drawn around these gauges (see Figure 3.6). Use this data to calculate the average depth over the area.

(3.3) In many cases rainfall is recorded on a cumulative basis but non-cumulative data are required for design purposes. Develop a general program which will convert a cumulative record to one giving individual rainfall data.

(3.4) Figure 3.5 shows the typical end result of a double mass analysis. Develop a routine for correcting records after the errors have been located from a plot such as that shown in Figure 3.5 and incorporate the routine into the program DOUBLEMASS. In Example 3.1 determine which gauge is faulty and adjust the record using your routine.

Chapter 4

Evaporation

ESSENTIAL THEORY

4.1 Introduction

Evaporation is the conversion of water from the liquid state into the gaseous phase and its diffusion into the atmosphere. Direct conversion of water from snow or ice into water vapour occurs by similar processes but is usually termed sublimation. Evaporation is normally thought of as taking place from water surfaces, e.g. reservoirs and lakes, but may be important when water is lost from wet soil or from land covered by vegetation. Relatively large amounts of water may be lost in this way. In the United States the annual evaporation varies from about 50 cm in Maine to over 200 cm in Southern California. This means that if lakes or reservoirs were not replenished by precipitation, groundwater inflow and rivers, etc., the lake level could drop up to 2 m over a twelve-month period. The quantities of water lost may be quite enormous and it has been estimated, for example, that the annual evaporation loss from lake Mead is approximately equivalent to the total annual water supply for the city of New York.

Evaporation can occur only when the appropriate amount of energy is added to convert the water from its liquid state to its gaseous phase; approximately 600 calories per gram. This is obtained from direct solar radiation or from heat in the overlying air, the ground or in the water itself. There must also be a vapour pressure gradient between the water surface and the overlying air. Vapour pressure is the partial pressure exerted by water vapour in the atmosphere. Saturation vapour pressure is the vapour pressure of a saturated air mass. This rises with temperature because warm air can hold a greater concentration of water vapour than cold air. If the air is not saturated and cooling occurs, a temperature will be reached at which the air will become saturated. This temperature is called the dew point. Any further cooling causes condensation. The relative humidity is the ratio of the actual vapour pressure to the saturation vapour pressure and is expressed as a percentage. It is a measure of the amount of moisture in a given space relative to the amount of moisture which could be held if the air were saturated.

Relative humidity can be measured using wet and dry bulb thermometers. The wet bulb is covered in a small muslin bag soaked in water. Air is moved past the thermometers by fans or by whirling them at the end of a cord. Under these conditions, evaporation from the cover of the wet bulb draws heat from that thermometer causing its temperature to drop. The difference in temperature after steady state conditions are reached provides an indication of the humidity. If the surrounding air is saturated, then blowing air past the wet bulb causes no evaporation and hence no drop in temperature.

This is the condition when the relative humidity is 100%. If the air is not saturated, the heat loss causes the temperature of the wet bulb to drop to a temperature at which the air is saturated. Psychrometric tables based on the two temperatures can be used to obtain the relative humidity. Alternatively, an approximate value (within 0.6%) from $-25°C$ to $+35°C$ can be obtained from

$$f = 100 \left[\frac{(112 - 0.1\, T_a + T_d)}{(112 + 0.9\, T_a)} \right]^8 \tag{4.1}$$

where f = relative humidity, T_a = temperature of air bulb (°C), and T_d = dew point temperature (°C).

The existence of a vapour pressure gradient implies a relative humidity less than 100%. Obviously if the air above a reservoir is completely saturated with water no water can evaporate from the reservoir. Wind therefore plays a large part in the evaporation process because, as water evaporates from the reservoir, the air at the surface will become saturated. Wind is therefore necessary to remove the saturated air mass and replace it with dryer air. Temperature also plays a large part because, although the primary energy input consists of solar radiation, heat energy may be available from other sources provided the ambient temperature is sufficiently high. In fact, temperature has a double effect because as the temperature of the air overlying the water rises, the saturation vapour pressure also rises making it possible for the air to absorb greater quantities of water vapour.

4.2   Measurement of evaporation

Evaporation cannot be measured directly but indirect measurements are possible. These generally rely on the use of a storage equation balancing inflow, outflow and storage. The equation may be written

in terms of a water balance or an energy balance for the reservoir. By measuring all quantities involved other than evaporation it is possible to use the equation in order to calculate evaporation. The water balance method is perhaps the most straightforward. Here the storage equation is written in terms of the volumes entering or leaving the reservoir over a period of time.
Thus,

$$P = E + Q + D + \Delta S \qquad (4.2)$$

where P = precipitation, E = evaporation, Q = net surface flow out of the reservoir, D = subsurface drainage and ΔS = increase in storage. Of the various terms in this equation, precipitation may always be regarded as an inflow and evaporation as a flow of water out of the reservoir. The other terms may be positive or negative depending on the direction of flow. Evaporation from a lake may be obtained by measuring each of the other terms in Equation (4.2). Precipitation and surface flow may be measured fairly precisely but problems can arise with the measurement of subsurface drainage and with changes in storage. However, if the measurement occurs over a long enough period, for example one year, it is possible to arrange the time period of the measurement so that the water level in the reservoir is the same at the beginning and end of the time period. In that case the change in storage will be zero and this term can be dropped from Equation (4.2). Measurement of subsurface drainage is much more difficult and must normally be estimated indirectly from measurements of permeability and groundwater levels etc. Under ideal conditions it has been estimated that the method has an accuracy of about ±10%.

The application of the storage equation to an energy budget yields

$$R_N + H = LH + G + \Delta H \qquad (4.3)$$

where R_N = net radiation = solar radiation − energy radiated away from the water body − radiation reflected from the water body, H = heat transfer from air to water, LH = heat used in converting liquid to vapour, i.e. in evaporation, G = heat flux into ground and ΔH = heat stored in water and heat removed by any outflow.

Measurements of all terms in the equation prove difficult. Net radiation may be measured by radiometers, black surfaces maintained under standard conditions with temperatures measured by thermocouples. Alternatively net radiation may be estimated from empirical relationships. Heat flux into or out of the ground can be measured with the use of buried heat flux plates. Measurements of the temperature difference across the plate provide an estimate of the heat passing across the plate. Heat flux from the air is normally obtained from empirical relationships. The method is highly specialised and could, perhaps, be considered to represent a physicist's approach to the problem rather than that of a hydrologist.

The simplest field method of measuring evaporation is undoubtedly that based on the use of evaporation pans. These are relatively inexpensive and, because it is possible to control all variables other than evaporation in the water balance equation, i.e. Equation (4.2), a fairly direct measurement is obtained. Various types and sizes of pan are available. The standard pan for North America is 4 ft (1.2 m) in diameter and 10 in (25 cm) deep. It is set with its base 6 in (15 cm) above ground level and the water surface is maintained within 2-3 in (5-8 cm) of the top of the pan. British pans are 6 ft square (3.3 m^2) 2 ft (60 cm) deep and set so that the rim of the pan is 3 in (8 cm) above surrounding ground level.

Since the relatively small amount of water in a pan is much more exposed to energy input in the form of heat than is the large mass of water in a reservoir, the evaporation from a pan tends to be somewhat higher than that which would be experienced from a lake. Because of this, the pan evaporation must be multiplied by a coefficient. A typical US class A pan coefficient would be about 0.70 but the actual coefficient will vary with location and with the type of pan. Measurements have indicated a range of values from 0.65 to 1.12.

4.3   Estimation of evaporation

Various empirical equations, based on the flow of vapour away from the reservoir, are available. These have the general form

$$E = (e_s - e_a) f(u) \tag{4.4}$$

where E = evaporation, e_s = saturation vapour pressure at water surface temperature, e_a = vapour pressure of air, f = function of, and u = wind speed at a specified distance above the water surface.

Equation (4.4) is based on the principle that evaporation depends primarily on the vapour pressure gradient at the surface and on the wind speed. This gives a measure of the potential evaporation assuming that a plentiful supply of water is available, for example the evaporation from a reservoir or lake. The actual evaporation from a land surface may be considerably less than the potential evaporation.

This method of estimating the evaporation is known as a vapour flow method, mass transfer technique or aerodynamic method. Because the equation is empirical, the function of wind speed $f(u)$

must be determined for each particular locality and the equation developed is not normally transferable to other situations. For example, measurements of evaporation at Lake Hefner in the United States have yielded

$$E = 0.122\,(e_s - e_a)\,u \tag{4.5}$$

This gives the lake evaporation in mm/day for a vapour pressure measured in mbar, 2 m above the water surface and for a wind speed defined in m/s, 4 m above the surface. This is significantly different from a formula of the same type developed for the Ijsselmeer in Holland and given by

$$E = 0.345\,(e_s - e_a)\,(1 + 0.25u) \tag{4.6}$$

in which the vapour pressures are measured in millimetres of mercury and the wind velocity is specified in m/s at a height of 6 m above the surface.

By combining the heat budget approach with the vapour flow method Penman derived an equation which can be used to calculate evaporation based only on meteorological data. This is

$$E = \frac{\Delta}{\Delta + \gamma} H + \frac{\gamma}{\Delta + \gamma} E_a \tag{4.7}$$

in which E = evaporation in mm/day, Δ = slope of the curve of saturation vapour pressure versus temperature, γ = psychrometer constant ($= 0.66$ when T is in °C and e in millibars), E_a = evaporation over open water per unit time (mm/day) assuming surface temperature and air temperature are equal and H = net radiation exchange measured in equivalent millimetres of evaporation per day.

Penman's equation can be solved in the following manner. The slope of the saturation vapour pressure curve may be obtained by differentiating an equation which approximates to that curve. This gives

$$\Delta = (0.00815\,T_a + 0.8912)^7 \tag{4.8}$$

where T_a = air temperature in °C.
Taking $\gamma = 0.66$

$$\frac{\Delta}{\Delta + \gamma} = \left[1 + \frac{0.66}{(0.00815\,T_a + 0.8912)^7} \right]^{-1} \tag{4.9}$$

and

$$\frac{\gamma}{\Delta + \gamma} = 1 - \frac{\Delta}{\Delta + \gamma} \tag{4.10}$$

The net radiation exchange is given by

$$H = 7.14 \times 10^{-3} R_A + 5.26 \times 10^{-6} R_A (T_a + 17.8)^{1.87}$$

$$+ 3.94 \times 10^{-6} R_A{}^2 - 2.39 \times 10^{-9} R_A{}^2 (T_a - 7.2) - 1.02 \quad (4.11)$$

where R_A = daily solar radiation (cal/cm^2/day) assuming a transparent atmosphere with no clouds. This quantity depends on latitude and on the time of year. Tabulated data are available in a variety of sources.

Open water evaporation is estimated from

$$E_a = (e_s - e_a)^{0.88} (0.42 + 0.0029 V_p) \quad (4.12)$$

where V_p = wind speed in km/day at a height of 150 mm above the surface and

$$(e_s - e_a) = 33.86 \left\{ \begin{array}{c} (0.00738 \, T_a + 0.8072)^8 \\ -(0.00738 \, T_d + 0.8072)^8 \end{array} \right\} \quad (4.13)$$

where T_d = dew point temperature. It should be noted that Equation (4.12) is based on the assumption, stated earlier, that the air temperature and water surface temperature are identical. This would be true only in a situation where the lake is extremely shallow or where the evaporation is occurring from something like an evaporation pan. To apply the results to a normal lake would therefore require the evaporation, given by Equation (4.7), to be multiplied by a pan type coefficient.

4.4 Evapo-transpiration

Evaporation refers specifically to the transfer of water from lake and soil surfaces into the atmosphere. Considerable quantities of water may also, however, be taken up by vegetation. Plants absorb water through their root systems and diffuse it back into the atmosphere through leaves. This process is known as transpiration. The amounts of water transpired vary with the climatic conditions, the type of plant and the amount of water available. Transpiration occurs primarily in the daylight hours and can vary from as much as 180 l/day for a large tree to about 2 l/day for an agricultural, small, plant.

When the ground is covered with vegetation, it is very difficult, if not impossible to separate evaporation from transpiration. In such cases, the two processes are linked and referred to as evapo-transpiration. Calculations of evapo-transpiration are based on the assumption that a plentiful supply of water is available. The amount determined is therefore referred to as the potential evapo-transpiration. Data showing transpiration from specific plants are of importance for agricultural and horticultural reasons. When considered by itself and referred to specific types of crops, transpiration is known as consumptive use.

A number of different methods exist for the determination of evapo-transpiration. Working with short dense vegetation in the United States, Thornthwaite obtained

$$P_e = 1.6 \left(\frac{10t}{J}\right)^a \frac{DT}{360} \text{ cm per month} \tag{4.14}$$

where P_e = potential evaporation, T = average number of hours between sunrise and sunset in the month, t = average temperature during the month, D = number of days in the month and

$$a = 675 \times 10^{-9} J^3 - 771 \times 10^{-7} J^2 + 179 \times 10^{-4} J + 0.492 \tag{4.15}$$

where J = yearly heat index given by

$$J = \sum_{n=1}^{n=12} \left(\frac{t_n}{5}\right)^{1.54} \tag{4.16}$$

and t_n = average monthly temperature in °C of consecutive months of the year ($n = 1,2,3 . . .12$).

Seasonal consumptive use is very difficult, time consuming and expensive to measure. However, data obtained in the United States have been used to develop coefficients which can be used in suitable empirical equations to determine the consumptive use of a crop under known climatic conditions. A typical equation for monthly consumptive use is given by the Blaney-Criddle relationship which can be expressed as

$$U = \frac{ktp}{100} \tag{4.17}$$

where k = consumptive use coefficient (dependent on crop and varies from month to month depending on climate), t = mean monthly temperature (°F) p = number of daylight hours in month as a percentage of the total number of daylight hours in the year and U = monthly consumptive use (ins water).

Data on coefficients and numbers of daylight hours are available from a variety of sources. Daylight hours are dependent on latitude and consumptive use coefficients on the type of crop.

WORKED EXAMPLES

Example 4.1 EVAPOR: estimation of evaporation

Compute the daily evaporation from a lake in a location where the air temperature is 20.0°C. The dew point is 7.0°C. Daily solar radiation is 530 cal/cm^2/day and wind speed is 9 km/h.

The evaporation can be calculated from Penman's equation (4.7). The various parts of Penman's equation are given in Equations (4.8) to (4.13) and will be solved in that order.

```
10 CLS
20 REM PENMAN'S EVAPORATION ANALYSIS
30 PRINT "PENMAN'S EVAPORATION ANALYSIS"
40 PRINT "CLASS 'A' PAN  - Pan Coefficient = 0.7"
50 PRINT
60 PRINT "ENTER ACTUAL TEMPERATURE IN deg C"
70 INPUT T
80 PRINT "ENTER DEW POINT TEMPERATURE IN deg C"
90 INPUT D
100 PRINT "ENTER DAILY SOLAR RADIATION IN cal/sq cm/day"
110 INPUT S
120 PRINT "ENTER WIND SPEED 150 mm ABOVE PAN IN km/day"
130 INPUT W
140 A = ( .00815 # T + .8911999 ) ^ 7
150 B = ( 1 + ( .66 / A ) )
160 C = ( 1 - B )
170 F=(.00714 # S)+(5.26E-06 # S # ((T + 17.8) ^ 1.87))+(3.94E-06 # (S ^ 2))-(2.39E-09 # (S ^ 2) # ((T - 7.2) ^ 2 ))-1.02
180 G = 33.86 # ( (( .00738 # T + .8072 ) ^ 8 ) - (( .00738 # D + .8072 ) ^ 8 ) )
190 H = ( G ^ .88 ) # ( .42 + ( .0029 # W ) )
200 M = .7 # ( ( B # F ) + ( C # H ) )
210 CLS
220 PRINT : PRINT : PRINT : PRINT
230 PRINT "PENMAN'S EVAVAPORATION ANALYSIS"
240 PRINT " Class 'A' Pan  - Pan Coefficient = 0.7 "
250 PRINT
260 PRINT "AVERAGE ACTUAL TEMPERATURE IN deg C IS "; T
270 PRINT "AVERAGE DEWPOINT TEMPERATURE IN deg C IS "; D
280 PRINT "DAILY SOLAR RADIATION IN calories per centimeter squared per day IS "; S
290 PRINT "WIND SPEED 150 mm ABOVE PAN IN km/day IS "; W
300 PRINT
310 PRINT "LAKE EVAPORATION IS "; M ;" mm/day . "
320 PRINT : PRINT : PRINT : PRINT
330 PRINT "DO YOU WISH TO DO THIS PROGRAM AGAIN ( Y/N ) ? "
340 INPUT A$
350 IF A$ = "Y" THEN GOTO 10
360 END


RUN

PENMAN'S EVAPORATION ANALYSIS
CLASS 'A' PAN  - Pan Coefficient = 0.7

ENTER ACTUAL TEMPERATURE IN deg C
? 20
ENTER DEW POINT TEMPERATURE IN deg C
? 7
ENTER DAILY SOLAR RADIATION IN cal/sq cm/day
? 530
ENTER WIND SPEED 150 mm ABOVE PAN IN km/day
? 216

PENMAN'S EVAVAPORATION ANALYSIS
 Class 'A' Pan  - Pan Coefficient = 0.7

AVERAGE ACTUAL TEMPERATURE IN deg C IS  20
AVERAGE DEWPOINT TEMPERATURE IN deg C IS  7
DAILY SOLAR RADIATION IN calories per centimeter squared per day IS  530
WIND SPEED 150 mm ABOVE PAN IN km/day IS  216

LAKE EVAPORATION IS  3.09455 mm/day .

DO YOU WISH TO DO THIS PROGRAM AGAIN ( Y/N ) ?
? N
```

Program notes

(1) Lines 0-50 — print title.
(2) Lines 60-130 — enter data.
(3) Lines 140-190 — calculate component parts of Penman's equation using Equations (4.8)-(4.13).
(4) Line 200 — calculates evaporation using Equation (4.17).
(5) Line 210 — clears screen*.
(6) Lines 220-310 — print data sentences.
(7) Lines 320-360 — permit recalculation with other data.

Note: not all BASICS can provide upper and lower case print.

Example 4.2 CONSUM: determination of consumptive use

Determine the monthly consumptive use from June to August inclusive, of two crops having consumptive use coefficients of 0.84 and 0.65, growing in a locality having temperatures and daylight as given below:

Month	% daylight hours	Air temperature (°F)
June	9.70	70
July	9.88	72
August	9.33	75

A simple program will be written on the basis of Equation (4.17) to provide monthly consumptive use, U, for a variety of values of k, t and p.

```
10 REM CONSUMPTIVE USE OF CROPS (BLANEY-CRIDDLE)
20 DIM M$(12)
30 M$(1)="JAN":M$(2)="FEB":M$(3)="MAR":M$(4)="APR":M$(5)="MAY":M$(6)="JUN"
40 M$(7)="JUL":M$(8)="AUG":M$(9)="SEP":M$(10)="OCT":M$(11)="NOV":M$(12)="DEC"
50 CLS
60 PRINT. "CONSUMPTIVE USE OF CROPS"
70 PRINT "-----------------------"
80 PRINT
90 PRINT "ENTER MONTH AT START OF GROWING SEASON e.g. 'APR'"
100 INPUT S$
110 PRINT "ENTER MONTH AT END OF GROWING SEASON e.g. 'SEP'"
120 INPUT F$
130 FOR I = 1 TO 12
140 IF (S$ = M$(I) ) THEN S = I
150 IF (F$ = M$(I) ) THEN F = I
```

*This statement is system dependent. Line 240 is valid on the IBM personal. (Apple machines use 'Home').

```
160 NEXT I
170 A = 0
180 PRINT "ENTER MEAN MONTHLY TEMPERATURE (in deg F)"
190 PRINT "PROPORTION OF DAYLIGHT HOURS (% of year)"
200 PRINT "VALUE OF 'K' FOR EACH MONTH AS PROMPTED"
210 PRINT
220 FOR I = S TO F
230 PRINT "FOR THE MONTH OF "; M$(I)
240 PRINT "MEAN TEMP. (deg F)"
250 INPUT T
260 PRINT "PROPORTION OF DAYLIGHT HOURS (% of year)"
270 INPUT P
280 PRINT "VALUE OF 'K' FOR TYPE OF CROP AND MONTH"
290 INPUT K
300 A = A + ( T$P$K )/100
305 PRINT : PRINT : PRINT
310 NEXT I
320 PRINT "THE CONSUMPTIVE USE OF CROP IS "; A ;" INCHES"
330 END

RUN
```

CONSUMPTIVE USE OF CROPS

ENTER MONTH AT START OF GROWING SEASON e.g. 'APR'
? JUN
ENTER MONTH AT END OF GROWING SEASON e.g. 'SEP'
? AUG
ENTER MEAN MONTHLY TEMPERATURE (in deg F)
PROPORTION OF DAYLIGHT HOURS (% of year)
VALUE OF 'K' FOR EACH MONTH AS PROMPTED

FOR THE MONTH OF JUN
MEAN TEMP. (deg F)
? 70
PROPORTION OF DAYLIGHT HOURS (% of year)
? 9.7
VALUE OF 'K' FOR TYPE OF CROP AND MONTH
? .84

FOR THE MONTH OF JUL
MEAN TEMP. (deg F)
? 72
PROPORTION OF DAYLIGHT HOURS (% of year)
? 9.88
VALUE OF 'K' FOR TYPE OF CROP AND MONTH
? .84

FOR THE MONTH OF AUG
MEAN TEMP. (deg F)
? 75
PROPORTION OF DAYLIGHT HOURS (% of year)
? 9.33
VALUE OF 'K' FOR TYPE OF CROP AND MONTH
? .84

THE CONSUMPTIVE USE OF CROP IS 17.56 INCHES

CONSUMPTIVE USE OF CROPS

ENTER MONTH AT START OF GROWING SEASON e.g. 'APR'
? JUN
ENTER MONTH AT END OF GROWING SEASON e.g. 'SEP'
? AUG
ENTER MEAN MONTHLY TEMPERATURE (in deg F)
PROPORTION OF DAYLIGHT HOURS (% of year)
VALUE OF 'K' FOR EACH MONTH AS PROMPTED

FOR THE MONTH OF JUN
MEAN TEMP. (deg F)
? 70
PROPORTION OF DAYLIGHT HOURS (% of year)
? 9.7
VALUE OF 'K' FOR TYPE OF CROP AND MONTH
? .65

FOR THE MONTH OF JUL
MEAN TEMP. (deg F)
? 72
PROPORTION OF DAYLIGHT HOURS.(% of year)
? 9.88
VALUE OF 'K' FOR TYPE OF CROP AND MONTH
? .65

FOR THE MONTH OF AUG
MEAN TEMP. (deg F)
? 75
PROPORTION OF DAYLIGHT HOURS (% of year)
? 9.33
VALUE OF 'K' FOR TYPE OF CROP AND MONTH
? .65

THE CONSUMPTIVE USE OF CROP IS 13.59 INCHES

Program notes

(1) Lines 10-20 — reserve array space for 12 months of year.
(2) Lines 30-40 — define string variables for each month, January-December.
(3) Lines 50-80 — clear screen and print title.
(4) Lines 90-160 — define growing season.
(5) Lines 180-290 — enter data as prompted.
(6) Line 300 — calculates monthly consumptive use and summates over growing season.
(7) Lines 305-330 — print total consumptive use and halt execution.

Example 4.3 HUMID: determination of humidity

The data given below were abstracted from psychrometric tables and show relative humidity and dew point temperature (in brackets) as a function of temperature and wet bulb depression. Develop a program to compare these data with values calculated using Equation (4.1).

Air temperature (°C)	Wet bulb depression (°C)										
	2		4		6		8		10		12
0	63	(−6)	29	(−16)	—		—		—		—	
10	77	(6)	55	(1)	34	(−5)	15	(−17)	—		—	
20	83	(17)	66	(14)	51	(10)	37	(5)	24	(−1)	12	(−4)
30	86	(27)	73	(25)	61	(22)	51	(18)	39	(15)	30	(10)
40	88	(38)	77	(35)	67	(33)	57	(30)	48	(27)	40	(24)

The program will set up two arrays to hold the specified values of air temperature, T_a, and wet bulb temperature, T_d. These will be used in Equation (4.1) to calculate relative humidity over the desired range of values.

```
10 CLS
20 REM        ### HUMIDITY CALCULATIONS
30 REM     VALUES CALCULATED FROM EQUATION 4.1 ( see text )
40 :
50 :
60 DEF FNHU(X,Y) =  ( ( ( 112 - ( .1 # X ) + Y ) / ( 112 + ( .9 # X ) ) ) ^ 8 ) # 100
70 DIM TA(5),TW(30),AN(30)
80 TA(1)=0:TA(2)=10:TA(3)=20:TA(4)=30:TA(5)=40
90 FOR I = 1 TO 23
100 READ TW(I)
110 NEXT I
120 J = 1
130 FOR I = 1 TO 2
140 AN(J) = FNHU( TA(1),TW(I) )
```

```
150 J = J + 1
160 NEXT I
170 FOR I = 1 TO 4
180 AN(J) = FNHU( TA(2),TW(I+2) )
190 J = J + 1
200 NEXT I
210 FOR I = 1 TO 5
220 AN(J) = FNHU( TA(3), TW(I+6) )
230 J = J + 1
240 NEXT I
250 FOR I = 1 TO 6
260 AN(J) = FNHU( TA(4), TW(I+11) )
270 AN(J+6) = FNHU( TA(5), TW(I+17) )
280 J = J + 1
290 NEXT I
300 FOR I = 1 TO 23
310 T1 = AN(I) * 100
320 T2= CINT(T1)
330 AN(I) = T2/ 100
340 NEXT I
350 PRINT "FNHU(X,Y) = ( ( ( 112 - ( .1 * X ) + Y ) / ( 112 + ( .9 * X ) ) ) ^ 8 ) * 100"
360 PRINT "--------------------------------------------------------------------------------"
370 PRINT "AIR TEMPERATURE";TAB(40);"WET BULB DEPRESSION"
380 PRINT TAB(8);"deg C";TAB(50);"deg C"
390 PRINT
400 PRINT TAB(20);"2";TAB(30);"4";TAB(40);"6";TAB(50);"8";TAB(60);"10";TAB(70);"12"
410 PRINT
420 PRINT TAB(6);"0";TAB(18);AN(1);TAB(28);AN(2)
430 PRINT TAB(5);"10";TAB(18);AN(3);TAB(28);AN(4);TAB(38);AN(5);TAB(48);AN(6)
440 PRINT TAB(5);"20";TAB(18);AN(7);TAB(28);AN(8);TAB(38);AN(9);TAB(48);AN(10);TAB(58);AN(11)
450 T = 11
460 FOR I = 30 TO 40 STEP 10
470 PRINT TAB(4) I;
480 FOR J = 1 TO 6
490 A1 = ( 8 + ( J * 10 ) )
500 PRINT TAB( A1 );AN( T + J );
510 NEXT J
520 T = T + 6
530 NEXT I
540 END
550 DATA -6,-16,6,1,-5,-17
560 DATA 17,14,10,5,-1
570 DATA 27,25,22,18,15,10
580 DATA 38,35,33,30,27,24
```

FNHU(X,Y) = (((112 - (.1 * X) + Y) / (112 + (.9 * X))) ^ 8) * 100

AIR TEMPERATURE   WET BULB DEPRESSION
     deg C   deg C

	2	4	6	8	10	12
0	64.37	29.14				
10	76.42	53.88	34.69	13.27		
20	82.96	68.52	52.71	37.5	24.43	
30	83.97	74.6	62.24	48.56	40.11	28.86
40	99.69	75.96	67.87	57.14	47.93	40.04

Program notes

(1) Lines 0-50 — clear screen and describe equation.
(2) Line 60 — defines humidity function (Equation (4.1)).
(3) Lines 70-80 — define arrays and air temperatures.
(4) Lines 90-110 — input dew point temperatures.
(5) Lines 120-290 — calculate answers with humidity function.
(6) Lines 300-340 — round off output.
(7) Lines 350-390 — print equation and headers.
(8) Lines 400-540 — output and print answers.
(9) Lines 550-580 — Data statements.

PROBLEMS

(4.1) Using Thornthwaite's method, calculate the potential evapo-transpiration in July (average daylight hours per day = 15) in a locality where the average monthly temperatures (°F) are

Jan.	Feb.	Mar.	Apr.	May	June	July	Aug.	Sept.	Oct.	Nov.	Dec.
33	37	43	55	65	71	76	75	70	58	45	35

(4.2) It has been suggested that Equations (4.15) and (4.16) respectively can be simplified to

$$a = 0.016J + 0.5$$

and $J = 0.09 (t_n)^{1.5}$

Modify the program developed in Problem (4.1) to incorporate these changes and compare the results with those obtained using Equations (4.15) and (4.16).

(4.3) Using the program developed in Problem (4.1) make the necessary adjustments to give *annual* potential evapo-transpiration.

(4.4) Equations (4.5) and (4.6) provide estimates of lake evaporation in two different locations. Develop a program to calculate the errors which would be experienced if the Dutch formula were to be used in the United States. Run the program with typical values of vapour pressure and wind velocity to determine the variation in results.

Chapter 5

Frequency

ESSENTIAL THEORY

5.1 Introduction

Frequency may be defined as the number of times, within a certain period of time, a specified occurrence is equalled or exceeded. For example, if one hundred years of records of annual rainfall are examined, then the biggest annual rainfall will be equalled on one occasion in the hundred years and will have a frequency of 1/100. The second biggest annual rainfall would be equalled on one occasion and would be exceeded on one occasion so the frequency would be 2/100 = 1/50. The fifth largest rainfall, for example, would be equalled or exceeded on five occasions out of the hundred years and would have a frequency of 1/20. In general then, if the record is rearranged so that the largest event is ranked number 1, the second largest number 2 and so on, then the frequency of any event is given by

$$F = m/n \qquad (5.1)$$

where F = frequency, m = rank and n = total number of years. The frequency calculated using Equation (5.1) is known as the Californian frequency.

One of the difficulties of analysing a record for point frequencies is that the frequency of the largest events cannot be precisely determined. If one hundred years of records are available, then the largest event has a frequency of 1/100. However, if one hundred and one years of records were available, it is possible that the extra year might be as large as, or greater than the biggest event in the one hundred years. The frequency of that event would then change from 1/100 to 2/100 = 1/50. In frequency calculations using annual rainfall records over fifty-five years, it was shown that the two-year frequency rain calculated from ten years of these records could differ by over 30% from the value determined from the complete record. In an attempt to overcome the uncertainties relating to frequency of the extreme events, various other formula have been

Table 5.1 Methods of defining frequency

Method	Formula
California	m/n
Hazen	$(2m-1)/2n$
Weibull	$m/(n+1)$
Beard	$1-(0.5)^{1/n}$
Chegadayev	$(m-0.3)/(n+0.4)$
Blom	$(m-0.375)/(n+0.25)$
Tukey	$(3m-1)/(3n+1)$

Adapted from Veissman *et al.*, 'Introduction to hydrology', In text Ed. Pub. New York.

developed (see Table 5.1). Of these the most common are the Weibull formula, also sometimes known as the Gumbel formula

$$F = m/(n+1) \tag{5.2}$$

and the Hazen formula

$$F = (2m-1)/2n \tag{5.3}$$

In calculating point frequencies, the Weibull formula is the most commonly used.

5.2 Recurrence interval

The recurrence interval, also called the return period, is the reciprocal of the frequency. For example, if in one hundred years of records an event is equalled or exceeded on ten occasions, the frequency will be 1/10 and the recurrence interval is 10 years. This does not mean however that if the event occurred last year it will not occur for another 10 years. There is no implication that the event is cyclical and to speak of an event as having a 10 year return period means only that over a long period of time it is likely to recur on average once in ten years. Over a thousand year period a ten year event might occur one hundred times but there is no suggestion that

Table 5.2 Return periods for design purposes

Type of structure	Return period (years)
Large bridge on freeway	200
Bridges on arterial roads	100
Bridges on local, secondary roads	50
Minor culverts	5–10
Storm water inlets	1–2

each occurrance would be separated by ten years. In fact, if each year is considered separately and independently, there is a 10% chance that the ten year event will occur during the one year period.

The recurrence interval of a particular event whether rainfall or flood, is of obvious importance for design considerations. Large return periods would be used for the design of structures where failure would result in excessive costs or loss of human life. In minor structures, however, where failure will result in only slight damage and where the cost of repair is minimal, a relatively short return period can be used. Table 5.2 indicates some return periods commonly used for design purposes.

5.3   Probability and risk

When a record of past events is analysed, Equations (5.1) to (5.3) may be used to define the frequency with which particular events have occurred in the past. For design purposes, however, it is important to be able to use that information for future predictions. In that sense the frequency of occurrence, as used for past records, provides a guide to the probability with which any particular event is likely to occur in the future. If the next hundred years is statistically identical to the last hundred, then the probability that the one hundred year event will occur in any of the next hundred years is 1/100. This average probability would be useful in designing a structure which had virtually unlimited life. However, it is also important to be able to estimate the probability with which a given event will occur over a specified period of time. This is called the risk. The probability R (risk) that an event with a return period T will occur at least once in a period of N successive years is given by

$$R = 1 - (1 - 1/T)^N \qquad (5.4)$$

The extrapolation of probabilities to return periods longer than the length of the record on which the calculations are based is not uncommon. Statistical methods of plotting frequency have been developed in attempts to obtain a straight line plot so that extrapolation can be carried out with some degree of safety. Although extrapolation of a straight line plot is advantageous in extending the record, it must be remembered that it is based on the assumption that the future will be statistically identical to the past. There can be little justification therefore in developing a one thousand year event from a record which is only, for example, ten years long.

5.4   Intensity-duration-frequency

The combination of intensity of rainfall with duration of the storm and the frequency of occurrence is particularly important for design calculations. Figure 5.1 shows an idealised catchment and a typical intensity duration curve. The design storm for the catchment is the one which gives the largest runoff. This depends on the intensity of

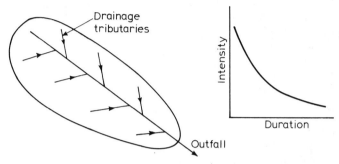

Figure 5.1   Idealised catchment and intensity-duration diagram

the storm and the area of the catchment contributing to the outflow. If the time taken for water to run from the most extreme point on the catchment to the outfall is T_c, the time of concentration, it is possible to consider the resulting flow at the outfall depending on the relationship between T_c and storm duration.

(1)   Storm duration $> T_c$: when the rain starts, the flow at the outfall will originate initially from the part of the catchment near the outfall. As time passes, more and more of the catchment contributes until water is running off the whole catchment. At that point, the flow becomes constant and is given by the product of the intensity and the area of the catchment.

(2)   Duration $< T_c$: when the duration of the storm is less than the time of concentration, the rain stops before water from the most extreme part of the catchment has arrived at the outfall. This means that at no one time does the whole of the catchment contribute to the flow.

(3)   Storm duration $= T_c$: under these conditions the rain stops at the instant when water from the most remote part of the catchment arrives at the outfall. The flow therefore begins to diminish as soon as it rises to a maximum.

These conditions are illustrated in Figure 5.2 which was drawn assuming constant intensity for simplicity. However, the intensity of the storm varies with duration such that high intensities correspond

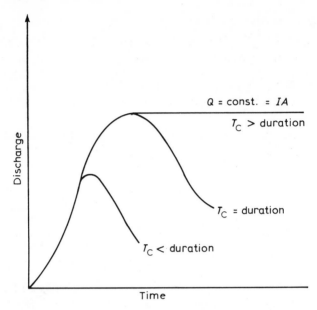

Figure 5.2   Catchment hydrographs assuming constant intensity

to short durations and low intensities to large durations. In condition 1 therefore, although the whole of the catchment contributes, the intensity of the storm would be relatively low. In condition 2 the intensity is high but the rain stops before the whole of the catchment contributes to the outfall. The optimum condition for design purposes must be such that the whole of the catchment contributes to the outfall at the maximum possible intensity and this will be obtained when the time of concentration is equal to the duration of the storm.

Intensity-duration-frequency diagrams are used to determine the intensity of the design storm. Frequency of occurrence is chosen depending on the importance of the structure as indicated earlier, and in Table 5.2, and the duration is set equal to the time of concentration for the particular catchment.

The combination of frequency with intensity duration curves was introduced earlier in Chapter 3, but the method in which Figure 3.8 was produced was illustrative rather than practical. In practice the construction of an intensity-duration frequency diagram would start by surveying the rainfall records so as to identify significant rainfall events with a fixed duration, for example, 10 min., 20 min., etc. These events would then be analysed for frequency of occurrence to develop one frequency-intensity curve for each of the specified durations. This data would then be re-analysed to provide a plot of intensity against duration for different frequencies.

WORKED EXAMPLES

Example 5.1   FREQUENCY: analysis of point frequency

The data below give annual rainfalls at a station over a time period of fifteen years. Write a program relating rainfall magnitude to frequency and develop it so that the Californian, Hazen or Weibull frequency can be calculated.

Year	1	2	3	4	5	6	7	8	9	10	11	12	13	14	15
Annual rainfall (mm)	250	300	500	480	750	600	550	700	400	350	610	550	700	450	680

A program will be set up so that the data can be read from DATA statements. A choice will then be made regarding the formula to be used, i.e. Equation (5.1), (5.2) or (5.3). Following that decision the data will be ranked using a bubble-sort technique and the frequency calculated according to the appropriate formula.

```
10 CLS
20 PRINT "FREQUENCY AND RETURN PERIOD ANALYSIS"
30 PRINT
40 PRINT "ENTER NUMBER OF OBSERVATIONS"
50 INPUT N
60 DIM R(N+1),S(N),P(N),T(N)
70 PRINT
80 FOR I = 1 TO N
90 READ R(I)
95 PRINT "OBSERVATION NO."; I ;" IS "; R(I)
100 NEXT I
110 PRINT : PRINT
115 PRINT "PRESS RETURN KEY ( enter ) TO CONTINUE !";
120 INPUT A$
130 FOR I = 1 TO (N-1)
140 FOR J = (I+1) TO N
150 IF R(I) > R(J) THEN GOTO 190
160 X = R(I)
170 R(I) = R(J)
180 R(J) = X
190 NEXT J
200 NEXT I
210 I = 1
220 B = 1
```

```
230 R(N+1) = 0
240 IF R(I) = R(I+1) THEN GOTO 290
250 FOR J = 0 TO (B-1)
260 S(I-J) = I
270 NEXT J
280 GOTO 320
290 I = I + 1
300 B = B + 1
310 GOSUB 740
320 B = 1
330 I = I + 1
340 GOSUB 740
350 CLS : PRINT "FREQUENCY ANALYSIS "
360 PRINT : PRINT : PRINT "OPTIONS :": PRINT
370 PRINT 1,"CALIFORNIA ( Tr = n/m ) "
380 PRINT 2,"HAZEN   ( Tr = 2n/(2m-1) )"
390 PRINT 3,"WEIBULL  ( Tr = (n+1)/m )"
400 PRINT 4,"END APPLICATION "
410 PRINT : PRINT
415 INPUT "CHOICE "; C$
418 C1 = VAL(C$)
420 IF ( C1 < 1 OR C1 > 4 ) THEN PRINT "BAD CHOICE NUMBER !": FOR I = 1 TO 1000 : NEXT I : GOTO 350
430 ON C1 GOTO 470,480,490,440
440 PRINT "END ... CONFIRM (Y/N)";
450 INPUT A$
460 IF A$ = "Y" THEN END
465 GOTO 350
470 F1 = 1 : F$ = "CALIFORNIA METHOD ( Tr = n/m ) " : GOTO 500
480 F1 = 2 : F$ = "HAZEN METHOD ( Tr = 2n/(2m-1) ) " : GOTO 500
490 F1 = 3 : F$ = "WEIBULL METHOD ( Tr = (n+1)/m ) " : GOTO 500
500 FOR I = 1 TO N
510 IF F1 = 1 THEN T(I) = N / S(I)
520 IF F1 = 2 THEN T(I) = ( 2*N ) / (( 2 * S(I) ) - 1.)
530 IF F1 = 3 THEN T(I) = ( N+1 ) / S(I)
540 P(I) = ( 1 / T(I) ) * 100
550 NEXT I
560 PRINT
570 CLS
600 PRINT
610 PRINT F$
620 PRINT
630 PRINT "OBSERV.      RANKING       PROB. (%)     Tr. (Years)"
640 PRINT "-------      --------      ----------    -------------"
650 PRINT
660 FOR I = 1 TO N
670 PRINT R(I),S(I),P(I),T(I)
680 NEXT I
690 PRINT
700 PRINT "DO YOU WISH TO DO THIS PROGRAM AGAIN WITH SAME DATA ( Y/N ) ?"
710 INPUT A$
720 IF A$ = "Y" THEN GOTO 350
730 GOTO 440
740 IF I > N THEN GOTO 350
745 GOTO 240
750 RETURN
760 DATA 250,300,500,480,750,600,550,700,400,350,610,550,700,450,680
```

RUN

FREQUENCY AND RETURN PERIOD ANALYSIS

ENTER NUMBER OF OBSERVATIONS
? 15

```
OBSERVATION NO. 1  IS  250
OBSERVATION NO. 2  IS  300
OBSERVATION NO. 3  IS  500
OBSERVATION NO. 4  IS  480
OBSERVATION NO. 5  IS  750
OBSERVATION NO. 6  IS  600
OBSERVATION NO. 7  IS  550
OBSERVATION NO. 8  IS  700
OBSERVATION NO. 9  IS  400
OBSERVATION NO. 10  IS  350
OBSERVATION NO. 11  IS  610
OBSERVATION NO. 12  IS  550
OBSERVATION NO. 13  IS  700
OBSERVATION NO. 14  IS  450
OBSERVATION NO. 15  IS  680
```

PRESS RETURN KEY < enter > TO CONTINUE !?

FREQUENCY ANALYSIS

OPTIONS :

```
1        CALIFORNIA ( Tr = n/m )
2        HAZEN  ( Tr = 2n/(2m-1) )
3        WEIBULL  ( Tr = (n+1)/m )
4        END APPLICATION
```

CHOICE ? 1

CALIFORNIA METHOD (Tr = n/m)

OBSERV.	RANKING	PROB. (%)	Tr. (Years)
750	1	6.666666	15
700	3	20	5
700	3	20	5
680	4	26.66667	3.75
610	5	33.33334	3
600	6	40	2.5
550	8	53.33333	1.875
550	8	53.33333	1.875
500	9	60.00001	1.666667
480	10	66.66667	1.5
450	11	73.33334	1.363636
400	12	80	1.25
350	13	86.66666	1.153846
300	14	93.33334	1.071429
250	15	100	1

```
DO YOU WISH TO DO THIS PROGRAM AGAIN WITH SAME DATA ( Y/N ) ?
? Y
```

FREQUENCY ANALYSIS

OPTIONS :

1	CALIFORNIA (Tr = n/m)
2	HAZEN (Tr = 2n/(2m-1))
3	WEIBULL (Tr = (n+1)/m)
4	END APPLICATION

CHOICE ? 2

HAZEN METHOD (Tr = 2n/(2m-1))

OBSERV.	RANKING	PROB. (%)	Tr. (Years)
750	1	3.333333	30
700	3	16.66667	6
700	3	16.66667	6
680	4	23.33334	4.285714
610	5	30	3.333333
600	6	36.66667	2.727273
550	8	50	2
550	8	50	2
500	9	56.66667	1.764706
480	10	63.33333	1.578947
450	11	70	1.428572
400	12	76.66667	1.304348
350	13	83.33333	1.2
300	14	90	1.111111
250	15	96.66667	1.034483

```
DO YOU WISH TO DO THIS PROGRAM AGAIN WITH SAME DATA ( Y/N ) ?
? Y
```

FREQUENCY ANALYSIS

OPTIONS :

1	CALIFORNIA (Tr = n/m)
2	HAZEN (Tr = 2n/(2m-1))
3	WEIBULL (Tr = (n+1)/m)
4	END APPLICATION

CHOICE ? 3

WEIBULL METHOD (Tr = (n+1)/m)

OBSERV.	RANKING	PROB. (%)	Tr. (Years)
750	1	6.25	16
700	3	18.75	5.333334
700	3	18.75	5.333334

680	4	25	4
610	5	31.25	3.2
600	6	37.5	2.666667
550	8	50	2
550	8	50	2
500	9	56.25	1.777778
480	10	62.5	1.6
450	11	68.75	1.454546
400	12	75	1.333333
350	13	81.25001	1.230769
300	14	87.50001	1.142857
250	15	93.75001	1.066667

```
DO YOU WISH TO DO THIS PROGRAM AGAIN WITH SAME DATA ( Y/N ) ?
? N

END ... CONFIRM (Y/N)? Y
Ok
```

Program notes

(1) Lines 0-30 — print title.

(2) Lines 40-70 — enter number of observations and reserve space for arrays.

(3) Lines 80-120 — read all data from DATA statement and output copy of data to screen.

(4) Lines 130-340 — bubble sort analysis to rank data.

(5) Lines 350-400 — define choice of four actions: three frequency calculations or stop.

(6) Lines 410-490—permit choice of above option and specify string variable (flag $) describing choice.

(7) Lines 500-550 — calculate frequency.

(8) Lines 560-570 — clear screen.

(9) Lines 600-650 — print titles.

(10) Lines 660-690 — print data.

(11) Lines 700-730 — permit recalculation using other frequency definition or other data.

(12) Lines 740-750 — subroutine test to determine if bubble sort is completed.

(13) Line 776 — DATA statement.

Example 5.2 FREQ: practical analysis of point frequency

For professional work involving large quantities of data the system for entering and handling information used in the previous program is unwieldy and inappropriate. An alternative is to record the data in permanent files and to use these as the data source. The following programs show how the files might be developed and used.

```
10 DIM OBS(300),RK(300),PROB(300),AN$(300),AN(300),FREQ(300)
20 ON ERROR GOTO 1840
30 :
40 :
50 REM                    ### MAIN MENU ###
60 CLS
70 PRINT : PRINT "FREQUENCY ANALYSIS"
80 PRINT : PRINT : PRINT "OPTIONS : " : PRINT
90 PRINT 1,"INPUT DATA ROUTINE "
100 PRINT 2,"PRINT DATA "
110 PRINT 3,"CALIFORNIA ( Tr = n/m )"
120 PRINT 4,"HAZEN ( Tr = 2n/(2m-1) )"
130 PRINT 5,"WEIBULL ( Tr = (n+1)/m )"
140 PRINT 6,"END APPLICATION"
150 PRINT : PRINT : INPUT "CHOICE" ; CHOICE$ : CHOICE = VAL(CHOICE$)
160 IF (CHOICE < 1) OR (CHOICE > 6) THEN PRINT "BAD CHOICE NUMBER" :FOR I = 1 TO 1000 : NEXT I : GOTO 60
170 ON CHOICE GOTO 230,470,850,860,870,190
180 GOTO 150
190 PRINT "END ... CONFIRM ";:GOSUB 1750 : IF A$ = "Y" OR A$ = "y" THEN END ELSE GOTO 60
200 :
210 :
220 :
230 REM                    ### INPUT DATA ROUTINE ###
240 :
250 CLS : PRINT "INPUT DATA ROUTINE " : PRINT
260 PRINT "DATA FILES AS FOLLOWS : "
270 FILES "?????FR.DAT"
280 PRINT : PRINT : PRINT
290 PRINT : PRINT "PLEASE ENTER FIVE CHARACTER DESIGNATION FOR DATA FILE "
300 PRINT "WARNING ! DO NOT REUSE FILE DESIGNATOR OR DATA WILL BE WRITTEN OVER !"
310 INPUT "FILE DESIGNATION (e.g. 'ABC12') ";M$
320 F$ = M$ + "FR.DAT"
330 OPEN "O",#1,F$
340 N = 1
350 KEY OFF
360 CLS : LOCATE 25,1 : PRINT "ENTER DATA ONE OBSERVATION AT A TIME <ENTER AN $> TO END DATA ENTRY " : LOCATE 1,1
370 PRINT "OBSERVATION "; N ;" IS : " ;
380 INPUT T$
390 IF T$ = "$" THEN CLOSE 1 : GOTO 60
400 PRINT#1,T$
410 N = N + 1
420 GOTO 370
430 :
440 :
450 :
460 :
470 REM                    ### PRINT DATA FILE ###
480 :
490 CLS : J = 1
500 PRINT "DATA FILES AS FOLLOWS : "
510 FILES "?????FR.DAT"
520 PRINT : PRINT : PRINT
530 INPUT "ENTER FIVE CHARACTER DESIGNATION e.g. 'ABC12' "; M$
540 F$ = M$ + "FR.DAT"
550 OPEN "I",#1,F$
560 IF EOF(1) THEN CLOSE 1 : GOTO 610
570 INPUT#1,AN$(J)
580 J = J + 1
590 GOTO 560
600 CLS
610 L = 1
620 CLS : PRINT : PRINT : PRINT
630 LPRINT : LPRINT : LPRINT
640 PRINT "DATA FILE : "; F$
650 LPRINT "DATA FILE : "; F$
660 PRINT "------------------------": PRINT : PRINT
670 LPRINT "------------------------": LPRINT : LPRINT
680 FOR I = 1 TO J
690 AN(I) = VAL(AN$(I))
700 IF AN(I) = 0 THEN GOTO 770
710 T = L * 8
720 IF T = 80 THEN L = 1 : T = L * 8
```

```
730 PRINT TAB(T) AN(I);
740 LPRINT TAB(T) AN(I);
750 L = L + 1
760 NEXT I
770 PRINT
780 LPRINT
790 GOSUB 1810
800 :
810 :
820 :
830 REM                  ### SUBROUTINE CONTROL AREA ###
840 :
850 FLAG = 10 : FLAG$ = "CALIFORNIA METHOD  ( Tr = n/m )" : GOTO 930
860 FLAG = 20 : FLAG$ = "HAZEN METHOD  ( Tr = 2n / (2m-1) )": GOTO 930
870 FLAG = 30 : FLAG$ = "WEIBULL METHOD  ( Tr = (n+1) / m )": GOTO 930
880 :
890 :
900 :
910 REM                  ### SUBROUTINE ORDER ###
920 :
930 GOSUB 1000
940 GOSUB 1130
950 GOSUB 1410
960 GOSUB 1510
970 :
980 :
990 :
1000 REM            ### FILE SELECTION SUBROUTINE ###
1010 :
1020 PRINT : PRINT : PRINT "DATA FILES AS FOLLOWS :"
1030 FILES "?????FR.DAT"
1040 PRINT : PRINT : PRINT
1050 PRINT "PLEASE ENTER FIVE CHARACTER DESIGNATION FOR DATA FILE "
1060 INPUT "FILE DESIGNATION (e.g. 'ABCDE') "; M$
1070 F$ = M$ + "FR.DAT"
1080 OPEN "I",#1,F$
1090 RETURN
1100 :
1110 :
1120 :
1130 REM          ### SORT AND RANK OBSERVATIONS SUBROUTINE ###
1140 :
1150 N = 1
1160 IF EOF(1) THEN CLOSE : GOTO 1210
1170 INPUT#1,T$
1180 OBS(N) = VAL(T$)
1190 N = N + 1
1200 GOTO 1160
1210 FOR I = 1 TO N
1220 FOR J = 1 TO N-1
1230 IF OBS(J) < OBS(J+1) THEN SWAP OBS(J),OBS(J+1)
1240 NEXT J
1250 NEXT I
1260 I = 1 : R = 1 :OBS(N+1) = 0
1270 IF OBS(I) = OBS(I+1) THEN GOTO 1320
1280 FOR J = 0 TO R-1
1290 RK(I-J) = I
1300 NEXT J
1310 GOTO 1330
1320 R = R + 1 : I = I + 1 : IF I > N THEN RETURN ELSE GOTO 1270
1330 R = 1 : I = I + 1
1340 IF I > N THEN RETURN ELSE GOTO 1270
1350 :
1360 :
1370 :
1380 :
1390 REM    ### CALCULATE RETURN PERIOD AND FREQUENCY SUBROUTINE ###
1400 :
1410 FOR I = 1 TO N
1420 IF FLAG = 10 THEN FREQ(I) = N / RK(I)
1430 IF FLAG = 20 THEN FREQ(I) = ( 2 * N ) / ( ( 2 * RK(I) ) - 1 )
1440 IF FLAG = 30 THEN FREQ(I) = ( N + 1 ) / RK(I)
```

```
1450 PROB(I) = ( 1 / FREQ(I) ) * 100
1460 NEXT I
1470 RETURN
1480 :
1490 :
1500 :
1510 REM          *** OUTPUT RESULTS SUBROUTINE ***
1520 :
1530 CLS : PRINT FLAG$ : PRINT : PRINT
1540 LPRINT FLAG$ : LPRINT : LPRINT
1550 GOSUB 1650
1560 FOR I = 1 TO N-1
1570 PRINT OBS(I),RK(I),FREQ(I),PROB(I)
1580 LPRINT OBS(I),RK(I),FREQ(I),PROB(I)
1590 NEXT I
1600 PRINT : PRINT : PRINT
1610 GOSUB 1810
1620 :
1630 :
1640 :
1650 REM               *** OUTPUT TITLES SUBROUTINE ***
1660 :
1670 PRINT "OBSERV.     RANKING          Tr         PROB "
1680 LPRINT "OBSERV.     RANKING          Tr         PROB "
1690 PRINT     "-------     -------       --------    --------"
1700 LPRINT    "-------     -------       --------    --------"
1710 RETURN
1720 END
1730 :
1740 :
1750 REM               *** QUESTION SUBROUTINE ***
1760 :
1770 PRINT "... (Y/N) ";
1780 A$ = INKEY$ :IF A$ <> "Y" AND A$ <> "N" THEN 1780 ELSE RETURN
1790 :
1800 :
1810 REM               *** OPTION RETURN ***
1820 :
1830 PRINT : PRINT : PRINT "DO YOU WISH TO GO TO MAIN MENU ? ";:GOSUB 1750:IF A$ = "Y" OR A$ = "y" THEN GOTO 60 ELSE 190
1840 IF ERR = 27 THEN LOCATE 23,1 :PRINT "TURN PRINTER ON ! ";: RESUME
1850 IF ERR = 24 THEN LOCATE 23,1 :PRINT "TURN PRINTER ON ! ";: RESUME
1860 IF ERR = 53 AND ERL = 270 THEN PRINT "FILE NOT FOUND": RESUME NEXT
1870 IF ERR = 53 AND ERL = 510 THEN PRINT "NO FILES ON DISK ! GO TO MENU AND USE INPUT DATA ROUTINE <1>": RESUME 1880
1880 FOR I = 1 TO 2000 : NEXT I : GOTO 30
```

FREQUENCY ANALYSIS

OPTIONS :

1	INPUT DATA ROUTINE
2	PRINT DATA
3	CALIFORNIA ($Tr = n/m$)
4	HAZEN ($Tr = 2n/(2m-1)$)
5	WEIBULL ($Tr = (n+1)/m$)
6	END APPLICATION

CHOICE? 1

INPUT DATA ROUTINE

DATA FILES AS FOLLOWS :
FILE NOT FOUND

PLEASE ENTER FIVE CHARACTER DESIGNATION FOR DATA FILE
WARNING ! DO NOT REUSE FILE DESIGNATOR OR DATA WILL BE WRITTEN OVER !
FILE DESIGNATION (e.g. 'ABC12') ? ABCDE

OBSERVATION 1 IS : ? 292		OBSERVATION 4 IS : ? 244
OBSERVATION 2 IS : ? 231		OBSERVATION 5 IS : ? 284
OBSERVATION 3 IS : ? 230		OBSERVATION 6 IS : ? 208

```
OBSERVATION    7  IS : ? 237          OBSERVATION   52  IS : ? 95
OBSERVATION    8  IS : ? 205          OBSERVATION   53  IS : ? 227
OBSERVATION    9  IS : ? 171          OBSERVATION   54  IS : ? 478
OBSERVATION   10  IS : ? 339          OBSERVATION   55  IS : ? 438
OBSERVATION   11  IS : ? 300          OBSERVATION   56  IS : ? 247
OBSERVATION   12  IS : ? 173          OBSERVATION   57  IS : ? 369
OBSERVATION   13  IS : ? 789          OBSERVATION   58  IS : ? 410
OBSERVATION   14  IS : ? 202          OBSERVATION   59  IS : ? 384
OBSERVATION   15  IS : ? 351          OBSERVATION   60  IS : ? 298
OBSERVATION   16  IS : ? 171          OBSERVATION   61  IS : ? 457
OBSERVATION   17  IS : ? 262          OBSERVATION   62  IS : ? 115
OBSERVATION   18  IS : ? 533          OBSERVATION   63  IS : ? 261
OBSERVATION   19  IS : ? 200          OBSERVATION   64  IS : ? 257
OBSERVATION   20  IS : ? 162          OBSERVATION   65  IS : ? 714
OBSERVATION   21  IS : ? 386          OBSERVATION   66  IS : ? 227
OBSERVATION   22  IS : ? 516          OBSERVATION   67  IS : ? 299
OBSERVATION   23  IS : ? 229          OBSERVATION   68  IS : ? 324
OBSERVATION   24  IS : ? 249          OBSERVATION   69  IS : ? 385
OBSERVATION   25  IS : ? 220          OBSERVATION   70  IS : ? 377
OBSERVATION   26  IS : ? 376          OBSERVATION   71  IS : ? 263
OBSERVATION   27  IS : ? 204          OBSERVATION   72  IS : ? 231
OBSERVATION   28  IS : ? 231
OBSERVATION   29  IS : ? 395          OBSERVATION   63  IS : ? 261
OBSERVATION   30  IS : ? 367          OBSERVATION   64  IS : ? 257
OBSERVATION   31  IS : ? 255          OBSERVATION   65  IS : ? 714
OBSERVATION   32  IS : ? 256          OBSERVATION   66  IS : ? 227
OBSERVATION   33  IS : ? 585          OBSERVATION   67  IS : ? 299
OBSERVATION   34  IS : ? 373          OBSERVATION   68  IS : ? 324
OBSERVATION   35  IS : ? 327          OBSERVATION   69  IS : ? 385
OBSERVATION   36  IS : ? 351          OBSERVATION   70  IS : ? 377
OBSERVATION   37  IS : ? 334          OBSERVATION   71  IS : ? 263
OBSERVATION   38  IS : ? 251          OBSERVATION   72  IS : ? 231
OBSERVATION   39  IS : ? 240          OBSERVATION   73  IS : ? 453
OBSERVATION   40  IS : ? 198          OBSERVATION   74  IS : ? 316
OBSERVATION   41  IS : ? 231          OBSERVATION   75  IS : ? 314
OBSERVATION   42  IS : ? 298          OBSERVATION   76  IS : ? 317
OBSERVATION   43  IS : ? 522          OBSERVATION   77  IS : ? 375
OBSERVATION   44  IS : ? 370          OBSERVATION   78  IS : ? 308
OBSERVATION   45  IS : ? 375          OBSERVATION   79  IS : ? 456
OBSERVATION   46  IS : ? 526          OBSERVATION   80  IS : ? 344
OBSERVATION   47  IS : ? 235          OBSERVATION   81  IS : ? 286
OBSERVATION   48  IS : ? 552          OBSERVATION   82  IS : ? 369
OBSERVATION   49  IS : ? 228          OBSERVATION   83  IS : ? 113
OBSERVATION   50  IS : ? 274          OBSERVATION   84  IS : ? 324
OBSERVATION   51  IS : ? 478          OBSERVATION   85  IS : ? 313
                                      OBSERVATION   86  IS : ? *
ENTER DATA ONE OBSERVATION AT A TIME <ENTER AN *> TO END DATA ENTRY




FREQUENCY ANALYSIS


OPTIONS :

    1          INPUT DATA ROUTINE
    2          PRINT DATA
    3          CALIFORNIA ( Tr = n/a )
    4          HAZEN ( Tr = 2n/(2a-1) )
    5          WEIBULL ( Tr = (n+1)/a )
    6          END APPLICATION


CHOICE? 2

DATA FILES AS FOLLOWS :
ABCDEFR .DAT



ENTER FIVE CHARACTER DESIGNATION e.g. 'ABC12' ? ABCDE
```

DATA FILE : ABCDEFR.DAT

```
292   231   230   244   284   208   237   205   171
339   300   173   789   202   351   171   262   533
200   162   386   516   229   249   220   376   204
231   395   367   255   256   585   373   327   351
334   251   240   198   231   298   522   370   375
526   235   552   228   274   478    95   227   478
438   247   369   410   384   298   457   115   261
257   714   227   299   324   385   377   263   231
453   316   314   317   375   308   456   344   286
369   113   324   313
```

DO YOU WISH TO GO TO MAIN MENU ? ... (Y/N)

FREQUENCY ANALYSIS

OPTIONS :

1	INPUT DATA ROUTINE
2	PRINT DATA
3	CALIFORNIA (Tr = n/m)
4	HAZEN (Tr = 2n/(2m-1))
5	WEIBULL (Tr = (n+1)/m)
6	END APPLICATION

CHOICE? 3

DATA FILES AS FOLLOWS :
ABCDEFR .DAT

PLEASE ENTER FIVE CHARACTER DESIGNATION FOR DATA FILE
FILE DESIGNATION (e.g. 'ABCDE') ? ABCDE

CALIFORNIA METHOD   (Tr = n/m)

OBSERV.	RANKING	Tr	PROB					
789	1	86	1.162791		373	24	3.583333	27.90698
714	2	43	2.325581		370	25	3.44	29.06977
585	3	28.66667	3.488372		369	27	3.185185	31.39535
552	4	21.5	4.651163		369	27	3.185185	31.39535
533	5	17.2	5.813954		367	28	3.071429	32.55814
526	6	14.33333	6.976745		351	30	2.866667	34.88372
522	7	12.28571	8.139535		351	30	2.866667	34.88372
516	8	10.75	9.302325		344	31	2.774194	36.04651
478	10	8.6	11.62791		339	32	2.6875	37.2093
478	10	8.6	11.62791		334	33	2.606061	38.37209
457	11	7.818182	12.7907		327	34	2.529412	39.53489
456	12	7.166667	13.95349		324	36	2.388889	41.86047
453	13	6.615385	15.11628		324	36	2.388889	41.86047
438	14	6.142857	16.27907		317	37	2.324324	43.02326
410	15	5.733333	17.44186		316	38	2.263158	44.18605
395	16	5.375	18.60465		314	39	2.205128	45.34884
386	17	5.058824	19.76744		313	40	2.15	46.51163
385	18	4.777778	20.93023		308	41	2.097561	47.67442
384	19	4.526316	22.09302		300	42	2.047619	48.83721
377	20	4.3	23.25581		299	43	2	50
376	21	4.095238	24.41861		298	45	1.911111	52.32559
375	23	3.739131	26.74419		298	45	1.911111	52.32559
375	23	3.739131	26.74419		292	46	1.869565	53.48837

286	47	1.829787	54.65117	230	67	1.283582	77.90698
284	48	1.791667	55.81396	229	68	1.264706	79.06977
274	49	1.755102	56.97674	228	69	1.246377	80.23256
263	50	1.72	58.13954	227	71	1.211268	82.55815
262	51	1.686275	59.30232	227	71	1.211268	82.55815
261	52	1.653846	60.46512	220	72	1.194444	83.72093
257	53	1.622642	61.62791	208	73	1.178082	84.88372
256	54	1.592593	62.7907	205	74	1.162162	86.04651
255	55	1.563636	63.95349	204	75	1.146667	87.20931
251	56	1.535714	65.11628	202	76	1.131579	88.37209
249	57	1.508772	66.27907	200	77	1.116883	89.53488
247	58	1.482759	67.44186	198	78	1.102564	90.69768
244	59	1.457627	68.60465	173	79	1.088608	91.86046
240	60	1.433333	69.76744	171	81	1.061728	94.18605
237	61	1.409836	70.93023	171	81	1.061728	94.18605
235	62	1.387097	72.09302	162	82	1.048781	95.34884
231	66	1.30303	76.74419	115	83	1.036145	96.51162
231	66	1.30303	76.74419	113	84	1.02381	97.67441
231	66	1.30303	76.74419	95	85	1.011765	98.83721
231	66	1.30303	76.74419				

FREQUENCY ANALYSIS

OPTIONS :

1	INPUT DATA ROUTINE
2	PRINT DATA
3	CALIFORNIA (Tr = n/a)
4	HAZEN (Tr = 2n/(2a-1))
5	WEIBULL (Tr = (n+1)/a)
6	END APPLICATION

CHOICE? 4

DATA FILES AS FOLLOWS :
ABCDEFR .DAT

PLEASE ENTER FIVE CHARACTER DESIGNATION FOR DATA FILE
FILE DESIGNATION (e.g. 'ABCDE') ? ABCDE

HAZEN METHOD (Tr = 2n / (2a-1))

OBSERV.	RANKING	Tr	PROB				
789	1	172	.5813953	384	19	4.648649	21.51163
714	2	57.33333	1.744186	377	20	4.410257	22.67442
585	3	34.4	2.906977	376	21	4.195122	23.83721
552	4	24.57143	4.069768	375	23	3.822222	26.16279
533	5	19.11111	5.232558	375	23	3.822222	26.16279
526	6	15.63636	6.395349	373	24	3.659575	27.32558
522	7	13.23077	7.55814	370	25	3.510204	28.48837
516	8	11.46667	8.72093	369	27	3.245283	30.81395
478	10	9.052631	11.04651	369	27	3.245283	30.81395
478	10	9.052631	11.04651	367	28	3.127273	31.97674
457	11	8.190476	12.2093	351	30	2.915254	34.30233
456	12	7.478261	13.37209	351	30	2.915254	34.30233
453	13	6.88	14.53488	344	31	2.819672	35.46512
438	14	6.370371	15.69768	339	32	2.730159	36.62791
410	15	5.931035	16.86046	334	33	2.646154	37.7907
395	16	5.548387	18.02326	327	34	2.567164	38.95349
386	17	5.212121	19.18605	324	36	2.422535	41.27907
385	18	4.914286	20.34884	324	36	2.422535	41.27907

317	37	2.356165	42.44186	235	62	1.398374	71.51163
316	38	2.293333	43.60465	231	66	1.312977	76.16279
314	39	2.233766	44.76744	231	66	1.312977	76.16279
313	40	2.177215	45.93023	231	66	1.312977	76.16279
308	41	2.123457	47.09303	231	66	1.312977	76.16279
300	42	2.072289	48.25582	230	67	1.293233	77.32559
299	43	2.02353	49.4186	229	68	1.274074	78.48838
298	45	1.932584	51.74419	228	69	1.255475	79.65117
298	45	1.932584	51.74419	227	71	1.219858	81.97674
292	46	1.89011	52.90698	227	71	1.219858	81.97674
286	47	1.849462	54.06977	220	72	1.202797	83.13955
284	48	1.810526	55.23256	208	73	1.186207	84.30232
274	49	1.773196	56.39535	205	74	1.170068	85.46512
263	50	1.737374	57.55815	204	75	1.154363	86.62791
262	51	1.70297	58.72093	202	76	1.139073	87.79069
261	52	1.669903	59.88372	200	77	1.124183	88.95349
257	53	1.638095	61.04651	198	78	1.109677	90.11628
256	54	1.607477	62.2093	173	79	1.095541	91.27907
255	55	1.577982	63.37209	171	81	1.068323	93.60464
251	56	1.54955	64.53488	171	81	1.068323	93.60464
249	57	1.522124	65.69767	162	82	1.055215	94.76744
247	58	1.495652	66.86046	115	83	1.042424	95.93024
244	59	1.470086	68.02326	113	84	1.02994	97.09302
240	60	1.445378	69.18605	95	85	1.017751	98.25581
237	61	1.421488	70.34884				

FREQUENCY ANALYSIS

OPTIONS :

1	INPUT DATA ROUTINE
2	PRINT DATA
3	CALIFORNIA (Tr = n/a)
4	HAZEN (Tr = 2n/(2a-1))
5	WEIBULL (Tr = (n+1)/a)
6	END APPLICATION

CHOICE? 5

DATA FILES AS FOLLOWS :
ABCDEFR .DAT

PLEASE ENTER FIVE CHARACTER DESIGNATION FOR DATA FILE
FILE DESIGNATION (e.g. 'ABCDE') ? ABCDE

WEIBULL METHOD   (Tr = (n+1) / a)

OBSERV.	RANKING	Tr	PROB				
789	1	87	1.149425	395	16	5.4375	18.3908
714	2	43.5	2.298851	386	17	5.117647	19.54023
585	3	29	3.448276	385	18	4.833334	20.68966
552	4	21.75	4.597701	384	19	4.578948	21.83908
533	5	17.4	5.747127	377	20	4.35	22.98851
526	6	14.5	6.896552	376	21	4.142857	24.13793
522	7	12.42857	8.045976	375	23	3.782609	26.43678
516	8	10.875	9.195402	375	23	3.782609	26.43678
478	10	8.7	11.49425	373	24	3.625	27.58621
478	10	8.7	11.49425	370	25	3.48	28.73563
457	11	7.909091	12.64368	369	27	3.222222	31.03448
456	12	7.25	13.7931	369	27	3.222222	31.03448
453	13	6.692308	14.94253	367	28	3.107143	32.18391
438	14	6.214286	16.09195	351	30	2.9	34.48276
410	15	5.8	17.24138	351	30	2.9	34.48276

344	31	2.806452	35.63218	244	59	1.474576	67.8161
339	32	2.71875	36.78161	240	60	1.45	68.96551
334	33	2.636364	37.93104	237	61	1.42623	70.11495
327	34	2.558824	39.08046	235	62	1.403226	71.26436
324	36	2.416667	41.37931	231	66	1.318182	75.86208
324	36	2.416667	41.37931	231	66	1.318182	75.86208
317	37	2.351351	42.52874	231	66	1.318182	75.86208
316	38	2.289474	43.67816	231	66	1.318182	75.86208
314	39	2.230769	44.82759	230	67	1.298508	77.0115
313	40	2.175	45.97701	229	68	1.279412	78.16092
308	41	2.121951	47.12644	228	69	1.26087	79.31035
300	42	2.071429	48.27587	227	71	1.225352	81.6092
299	43	2.023256	49.42529	227	71	1.225352	81.6092
298	45	1.933333	51.72414	220	72	1.208333	82.75862
298	45	1.933333	51.72414	208	73	1.191781	83.90804
292	46	1.891304	52.87357	205	74	1.175676	85.05748
286	47	1.851064	54.02299	204	75	1.16	86.2069
284	48	1.8125	55.17241	202	76	1.144737	87.35631
274	49	1.77551	56.32184	200	77	1.12987	88.50574
263	50	1.74	57.47127	198	78	1.115385	89.65518
262	51	1.705882	58.62069	173	79	1.101266	90.8046
261	52	1.673077	59.77012	171	81	1.074074	93.10345
257	53	1.641509	60.91954	171	81	1.074074	93.10345
256	54	1.611111	62.06896	162	82	1.060976	94.25288
255	55	1.581818	63.21839	115	83	1.048193	95.4023
251	56	1.553572	64.36781	113	84	1.035714	96.55172
249	57	1.526316	65.51725	95	85	1.023529	97.70115
247	58	1.5	66.66667				

FREQUENCY ANALYSIS

OPTIONS :

1	INPUT DATA ROUTINE
2	PRINT DATA
3	CALIFORNIA (Tr = n/m)
4	HAZEN (Tr = 2n/(2m-1))
5	WEIBULL (Tr = (n+1)/m)
6	END APPLICATION

CHOICE? 6
END ... CONFIRM ... (Y/N)

Program notes

(1) Lines 0-40 —dimension arrays to 300 variable and set up error trapping routine.

(2) Lines 50-220 = MAIN MENU. Lines 70-150 set up screen with menu; line 160 test for incorrect selection; line 170 go to correct area of program; line 190 confirm end of program.

(3) Lines 230-460 = INPUT DATA ROUTINE. Lines 250-310 set up screen and input file name; lines 320-330 set file spec. and open file for output; lines 350-360 message at line 25; lines 370-390 input observations; lines 400-420 write to file; increment counter and return.

(4) Lines 470-820 = PRINT DATA FILE. Lines 490-530 set up screen and input file name; lines 540-550 set file spec. and open file for input; lines 560-600 test for end of file and input data to array;

lines 610-670 output file name to screen and printer; lines 680-760 loop to output file data; lines 770-790 option return (Gosub 1810).
(5) Lines 830-900 = SUBROUTINE CONTROL AREA. Set flag and string description for each of the options.
(6) Lines 910-990 = SUBROUTINE ORDER. Line 930: file selection subroutine; line 940: sort and rank observations subroutine; line 950: calculate return period and frequency subroutine; line 960: output results subroutine.
(7) Lines 1000-1120 = FILE SELECTION SUBROUTINE. Lines 1020-1060 set up screen and input file name; lines 1070-1090 set file spec. open files for input and return to subroutine order.
(8) Lines 1130-1380 = SORT AND RANK OBSERVATION SUBROUTINE. Lines 1150-1200 input observations to array OBS (N); lines 1210-1250 bubble sort (highest to lowest); lines 1260-1340 rank observations and track duplicate values and returns.
(9) Lines 1390-1500 = CALCULATE RETURN PERIOD AND FREQUENCY SUBROUTINE. Loop to calculate return period based on flag option with frequency calculated on return period.
(10) Lines 1510-1640 = OUTPUT RESULTS SUBROUTINE. Lines 1530-1540 clear screen and output option selected; line 1550 go to titles subroutine (Gosub 1650); lines 1560-1600 loop to print observation, rank, return period and probability; line 1610 option return (Gosub 1810).
(11) Lines 1650-1740 = OUTPUT TITLES SUBROUTINE. Print title headings on screen and printer.
(12) Lines 1750-1800 = QUESTION SUBROUTINE. Ask for yes and no answer.
(13) Lines 1810-1830 = OPTION RETURN. Ask if you wish to return to menu.
(14) Lines 1840-1880 = ERROR TRAPPING AREA.

Example 5.3 RISK: analysis of risk

Determine the return periods for events which have a 1, 5, 20, 50, 80, 95 and 99 % probability of occurring at least once during the life of projects designed for 10, 20, 30 and 40 years.

This is a basic statistical problem which can be solved by developing a program to manipulate Equation (5.4) for $1 < R < 99$ and $10 < N < 40$. Data should be printed in tabular form showing return period (T) as a function of risk (R) and life of project (N).

```
10 CLS
20 PRINT "RISK ANALYSIS"
30 PRINT "-------------"
40 PRINT : PRINT : PRINT
```

```
50 PRINT "DO YOU WISH TO MODIFY THE DATA STATEMENTS (Y/N) ";
60 INPUT A$ : IF A$ = "Y" THEN STOP
70 REM RISK ANALYSIS TABLE
80 REM READ DATA TO GET NUMBER OF FAILURE FACTORS
90 READ N
100 REM READ DATA TO GET NUMBER OF LIFE OF PROJECT FACTORS
110 READ M
120 REM SET UP ARRAYS
130 DIM R(N),L(M),R1(M,N)
140 REM SET UP INPUT LOOP
150 FOR I = 1 TO N
160 READ R(I)
170 NEXT I
180 FOR I = 1 TO M
190 READ L(I)
200 NEXT I
210 REM SET UP FOR CALCULATIONS
220 FOR I = 1 TO M
230 FOR J = 1 TO N
240 EP = 1 / L(I)
250 P = 1 - ( 1 - R(J) ) ^ EP
260 R1(I,J) = 1 / P
270 NEXT J
280 NEXT I
290 REM OUTPUT RESULTS
300 CLS
310 PRINT "   RETURN PERIOD REQUIRED FOR SPECIFIC RISK OF OCCURENCE WITHIN PROJECT LIFE "
320 PRINT "-----------------------------------------------------------------------------"
330 PRINT "EXPECTED LIFE"
340 PRINT "OF PROJECT (yrs)                    PERMISSIBLE RISK OF FAILURE"
350 PRINT "                 ----------------------------------------------------------"
360 PRINT TAB(16);
370 FOR I = 1 TO N
380 PRINT USING "####.##"; R(I);
390 NEXT I
400 PRINT "                 ----------------------------------------------------------"
410 FOR I = 1 TO M
420 PRINT TAB(7);L(I);TAB(16);
430 FOR J = 1 TO N
440 PRINT USING "#####.#";R1(I,J);
450 NEXT J
460 NEXT I
470 END
480 DATA 9
490 DATA 17
500 DATA 0.01,0.05,0.10,0.20,0.50,0.70,0.80,0.95,0.99
510 DATA 1,2,3,4,5,10,15,20,25,30,40,50,60,70,80,90,100
```

```
RUN

RISK ANALYSIS
-------------

DO YOU WISH TO MODIFY THE DATA STATEMENTS (Y/N) ? N
```

```
RETURN PERIOD REQUIRED FOR SPECIFIC RISK OF OCCURENCE WITHIN PROJECT LIFE
--------------------------------------------------------------------------
EXPECTED LIFE
OF PROJECT (yrs)                    PERMISSIBLE RISK OF FAILURE
                  -------------------------------------------------------------
              0.01    0.05   0.10   0.20   0.50   0.70   0.80   0.95   0.99
                  -------------------------------------------------------------
        1     100.0    20.0   10.0    5.0    2.0    1.4    1.3    1.1    1.0
        2     199.5    39.5   19.5    9.5    3.4    2.2    1.8    1.3    1.1
        3     299.0    59.0   29.0   14.0    4.8    3.0    2.4    1.6    1.3
        4     398.5    78.5   38.5   18.4    6.3    3.8    3.0    1.9    1.5
        5     498.0    98.0   48.0   22.9    7.7    4.7    3.6    2.2    1.7
       10     995.5   195.5   95.4   45.3   14.9    8.8    6.7    3.9    2.7
       15    1493.2   292.9  142.9   67.7   22.1   13.0    9.8    5.5    3.8
       20    1990.7   390.4  190.3   90.1   29.4   17.1   12.9    7.2    4.9
       25    2488.1   487.9  237.8  112.5   36.6   21.3   16.0    8.9    5.9
       30    2985.3   585.4  285.2  134.9   43.8   25.4   19.1   10.5    7.0
       40    3980.4   780.3  380.1  179.8   58.2   33.7   25.4   13.9    9.2
       50    4976.9   975.3  475.1  224.6   72.6   42.0   31.6   17.2   11.4
       60    5970.5  1170.3  570.0  269.4   87.1   50.3   37.8   20.5   13.5
       70    6967.3  1365.2  664.9  314.2  101.5   58.6   44.0   23.9   15.7
       80    7958.8  1560.2  759.8  359.0  115.9   66.9   50.2   27.2   17.9
       90    8952.6  1755.1  854.7  403.8  130.3   75.3   56.4   30.5   20.0
      100    9950.9  1949.9  949.6  448.6  144.8   83.6   62.6   33.9   22.2

RISK ANALYSIS
-------------
DO YOU WISH TO MODIFY THE DATA STATEMENTS (Y/N) ? Y
Break in 60
Ok
480 DATA 7
490 DATA 4
500 DATA 0.01,0.05,0.20,0.50,0.80,0.95,0.99
510 DATA 10,20,30,40
RUN
RISK ANALYSIS
-------------
DO YOU WISH TO MODIFY THE DATA STATEMENTS (Y/N) ? N

   RETURN PERIOD REQUIRED FOR SPECIFIC RISK OF OCCURENCE WITHIN PROJECT LIFE
--------------------------------------------------------------------------------
EXPECTED LIFE
OF PROJECT (yrs)                    PERMISSIBLE RISK OF FAILURE
                  -------------------------------------------------------------
              0.01    0.05   0.20   0.50   0.80   0.95   0.99
                  -------------------------------------------------------------
       10     995.5   195.5   45.3   14.9    6.7    3.9    2.7
       20    1990.7   390.4   90.1   29.4   12.9    7.2    4.9
       30    2985.3   585.4  134.9   43.8   19.1   10.5    7.0
       40    3980.4   780.3  179.8   58.2   25.4   13.9    9.2
```

Program notes

(1) Lines 0-70 — remark, print title and ask questions about DATA statements*.

*Statements 50 and 60 permit the DATA statements to be modified. When the user types Y (yes) following the question the program stops at line 60. The DATA statements may then be rewritten and execution restarted.

(2) Lines 80-110 — read basic data from DATA statements.
(3) Lines 120-130 — reserve array space.
(4) Lines 140-200 — read data from DATA statements.
(5) Lines 210-280 — calculate return period from Equation (5.4).
(6) Lines 290-400 — set up table and print headings.
(7) Lines 410-460 — Print life, risk and return periods.
(8) Line 470 — end of program.
(9) Lines 480-510 — DATA statements.

PROBLEMS

(5.1) Determine the risk that events having return periods given in Table 5.2 will occur at least once in periods of two years and five years. Use the larger value where a range is indicated.

(5.2) From the annual precipitation data given below, estimate the maximum rainfall that can be expected in 10, 20 and 30 years. Use the Hazen method of plotting and compare your results with those obtained from the Weibull formula.

Year	1	2	3	4	5	6	7	8	9	10	11	12	13	14	15	16
Annual rainfall (mm)	923	711	733	787	620	600	893	459	715	652	749	458	619	642	656	768

(5.3) The data below give maximum annual discharge in a river over a period of 30 years. Compare the magnitude of the two-year flood obtained from the full record with that obtained from any consecutive ten-year period.

Year	Flow (m³/s)	Year	Flow (m³/s)
1	1119	16	1017
2	1754	17	1944
3	2170	18	2043
4	1196	19	1493
5	799	20	878
6	1587	21	1156
7	1997	22	2043
8	1496	23	2306
9	1474	24	1300
10	1235	25	1246
11	1975	26	1790
12	1768	27	970
13	1405	28	1796
14	1669	29	1303
15	1694	30	1048

Chapter 6

Stream flow and surface runoff

ESSENTIAL THEORY

6.1 Factors affecting runoff

Water in stream channels originates from three distinct and different sources. Some water, but relatively little, results from precipitation falling directly on the stream surface. This is usually insignificant except for very large rivers or rivers connected to lakes with extensive surface areas. Precipitation which falls in the catchment area around the river may get into the stream channels in two ways. The water may percolate down through the soil and into the water table. If this is higher than the river level, water will flow towards the river (see Figure 6.1). The proportion of the river flow which originates from groundwater can vary substantially but changes in groundwater flow are extremely slow. Some water falling on the ground may run into the stream channels without percolating into the water table. Water originating in this fashion is called surface runoff. This is

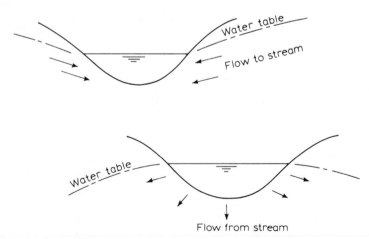

Figure 6.1 Drainage to and from stream channels

most easily seen on impermeable surfaces such as roads and parking areas but does also occur on pervious ground. Compared to groundwater flow surface runoff varies quickly. It is therefore the prime cause of floods in river channels.

Surface runoff occurs when rain falls with an intensity greater than the rate at which it is able to infiltrate the soil and penetrate into the water table. This rate is called the infiltration capacity. It depends on the permeability of the suface and on antecedent rainfall which is important because of its effect on the moisture level in the soil. Any increase in soil moisture will reduce the soil's capacity to absorb rain and will thus decrease the infiltration capacity. Surface runoff occurs, therefore, when the rainfall intensity exceeds the infiltration capacity. It is the excess rainfall which runs quickly into the stream channels and gives rise to floods. In arid countries with pervious soils, surface runoff may occur only with high intensity storms. It is therefore possible that, throughout most of the year, the stream channels will be completely dry and that sudden flash floods will occur following heavy rain storms.

The factors affecting runoff may be classified generally in two groups: climatic and physiographic. Climatic features include such things as type of precipitation, typical intensity and duration of storms, distribution of rainfall and direction of storm movement. For example, the type of precipitation, rain or snow, affects the time of the year in which the major runoff occurs. In temperate countries it is common for the maximum runoff to occur in the winter season. In countries having a continental climate, winter precipitation often takes the form of snow and maximum runoff does not occur until the snow melts, usually in late spring. As discussed previously, antecedent rainfall and soil moisture levels affect the runoff, and the rapidity with which storms follow one another is therefore of importance. In general all climatic conditions have some effect on runoff because the climatic conditions affect plant type and growth. These in turn affect transpiration and interception, etc. It is well known for example, that the runoff from a well-mulched forest is considerably less than that from an agricultural or an urban area.

The type of soil and the area of the catchment basin have obvious effects on the runoff because they affect the infiltration capacity and the volume of rainfall which is directed towards the stream channel. However, other physiographic features are also of importance. The slope of the basin and the sinuosity of the drainage channels have a large effect in determining the rate of runoff. The elevation of the basin determines, to some extent, the type of precipitation, rain or snow, and affects water losses and transpiration. Relatively fast runoff occurs in well-drained basins and the drainage network is therefore of importance. The orientation of the basin, whether facing north or south, affects the extent of the snowfall and the time of snow melt. It also affects plant growth and transpiration. The effects of cultivation have already been considered. Cultivation generally increases runoff.

6.2   The hydrograph

The hydrograph of a river or channel shows the variation in flow rate with respect to time (see Figure 6.2). As discussed in the previous

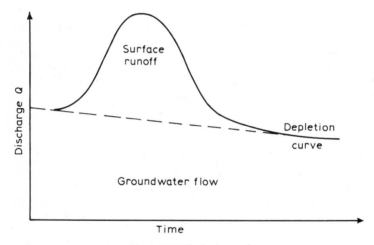

Figure 6.2   The hydrograph

section, the flow in a river originates from two main sources — surface runoff and groundwater flow. When rain falls and surface runoff occurs, the flow in the river rises. When the rain stops and surface runoff ceases, the flow decreases and, provided there is an extended period without rainfall, the flow in the river will decrease until comprised of groundwater alone. As the level of the water table falls, the flow of groundwater decreases and, under these conditions the hydrograph represents a curve showing groundwater depletion. As illustrated in Figure 6.3 groundwater depletion curves can be particularly striking for tropical rivers in which there is little or no rain during an extended dry season.

The separation of the hydrograph into surface runoff and groundwater flow can be done using master depletion curves or, in their absence, by simpler empirical techniques. It is probable that the separation should follow the lines shown in Figure 6.4, but

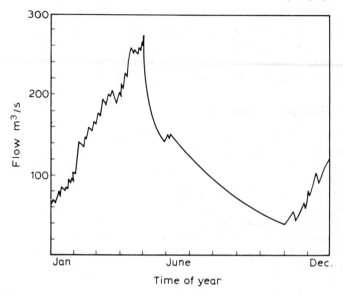

Figure 6.3   Typical annual flow variation in some tropical rivers

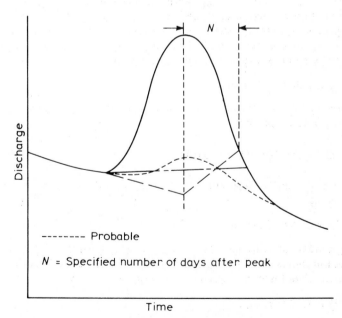

Figure 6.4   Various methods of separating base flow

somewhat arbitrary methods using straight separation lines may often be used with few inaccuracies. The area under the curve of surface runoff is equal to the volume of surface runoff. This may be correlated with excess rainfall, the volume being given by the product of excess rainfall and area of catchment.

6.3   Flow measurement by weirs

In small streams and channels, flow measurement can be easily accomplished using inexpensive sharp crested weir structures. In many cases these can be constructed from plywood and temporary, portable arrangements may be designed for use in very small channels. The weir is essentially an obstruction built across the channel and designed to ensure that the whole of the flow is directed through a notch, usually located centrally within the weir. Care must be taken to minimise or eliminate leakage around the weir.

One of the most common weirs for small flows is the triangular, or V-notch, weir in which

$$Q = \frac{8}{15} C_d \tan \frac{\theta}{2} \ (2g)^{1/2} H^{5/2} \qquad (6.1)$$

where Q = discharge, C_d = discharge coefficient, θ = notch angle, g = gravitational acceleration and H = head over notch.

The most commonly used V-notch is that having a 90° notch angle but, if the flows are very small, a 45° or perhaps 22½° V-notch weir will give greater accuracy. For a 90° notch Equation (6.1) reduces to

$$Q = 2.362\, C_d^{\ 5/2}\, \text{m}^3/\text{sec} \qquad (6.2)$$

The rectangular weir is used in a variety of circumstances. When side contractions are present

$$Q = \frac{2}{3} C_d (2g)^{1/2} (L - 0.2H) H^{3/2} \qquad (6.3)$$

where L = length of weir crest. if there are no end contractions, this reduces to

$$Q = 2.95\, C_d L H^{3/2}\, \text{m}^3/\text{sec} \qquad (6.4)$$

Trapezoidal or Cipoletti weirs are used extensively on small irrigation canals. The relationship between discharge and head over the weir when losses are taken into account is given by

$$Q = 1.87\, L H^{3/2}\, \text{m}^3/\text{sec} \qquad (6.5)$$

(for $\theta/2 \approx 14°$).

Flumes are specially shaped open channel flow sections in which the discharge is directly related to the depth. Generally they involve a constriction of the flow passage causing an increase in velocity and the occurrence of critical flow. As such they offer distinct advantages to the use of weirs in situations where the channel carries sediment or solid material, for example in sewer channels. The most commonly used flume is the parshall flume. This consists essentially of a short converging area, a throat with parallel walls in which the elevation of the bottom drops and a diverging area with a rising floor back to the width of the main channel. For a given geometry the discharge is directly related to the upstream depth. Tabulated data are available in a variety of sources.

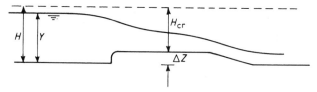

Figure 6.5 Critical flow weir

In large rivers, the flow is most commonly measured with the aid of a broad-crested weir constructed across the river. If the channel is rectangular, and if it is assumed that there are no energy losses, then the depth and discharge for the configuration shown in Figure 6.5 are related by

$$H = y + \frac{q^2}{2gy^2} = (H_{cr} + \Delta Z) = \frac{3}{2}(q^2/g)^{1/3} + \Delta Z \qquad (6.6)$$

where H = head over weir, y = upstream depth, q = flow per unit width, H_{cr} = critical specific energy and ΔZ = height of weir. Equation (6.6) shows that the depth upstream of the weir is uniquely related to the discharge in the channel. Thus, although Equation (6.6) can only be used for calculation in special circumstances, the principle that the depth and discharge are uniquely related may be used for practical flow measurement.

If the discharge is physically measured for a variety of depths, then it is possible to produce a stage discharge diagram similar to that shown in Figure 6.6. In order to cover a full range of flows, it would be necessary to measure flows throughout the entire year, some being measured in the dry season and some in the wet season. However, once the stage discharge diagram has been prepared, measurements of flow may be obtained as often as required, perhaps on a daily basis, simply

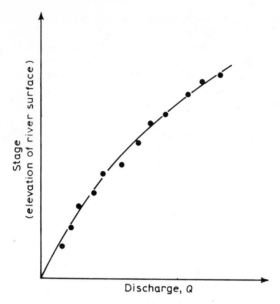

Figure 6.6 Stage discharge diagram

by measuring the elevation of the water surface. This may be accomplished using a simple staff gauge rigidly fixed to the bank of the river or by the use of recording gauges utilising a float and rotating drum for continuous measurement.

Measurements of flow for the purpose of developing a stage discharge diagram are normally accomplished by measuring the velocity profile of the river. As shown in Figure 6.7, the river would be divided into a number of vertical sections. The cross section would be

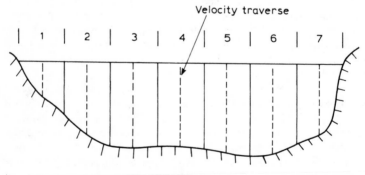

Figure 6.7 River gauging

plotted to determine the area and the average velocity in each section would be measured by a current meter. A variety of methods of obtaining average velocity is available. In large rivers a full vertical traverse would be required but in smaller rivers it is possible, with reasonable accuracy, to take the average of the velocities measured at 0.2 and 0.8 of the depth. This gives the mean velocity to within ±2%. A single point measurement at 0.6 of the depth is still reasonably accurate for small rivers and provides a measure of the mean velocity to within 5%. Having evaluated the average velocity in each section, the discharge through each section is obtained from the product of velocity and area. These are summated to provide a measure of the total discharge.

6.4   Empirical equations for prediction of runoff

For a region in which the climatic condition is generally constant, the runoff to be expected from a storm will depend on the physiographic features of the catchment area and the particulars of the storm causing runoff. Thus it should be possible to correlate the runoff with these features in an equation having the general form

$$Q = K I^a D^b A^c S^d S_i^e V^f \dots \tag{6.7}$$

where I = intensity, D = duration, A = area, S = slops, S_i = sinuosity, V = vegetation factor, etc. and K, a, b, c, etc. are constants. Values of the constants can be obtained by an analysis of runoff in various different catchments of the region. The variables included in the equation would be measured and a multiple regression technique could be used to provide values of the unknowns. The development of such an equation requires extensive records, a good correlation between runoff and the rainfall causing it and some commonsense. The indices a, b, c, etc. indicate the importance of the particular parameter and it would make little sense if, for example, the area of the catchment were raised to a very small power (e.g. $A^{0.002} \approx 1.0$). Regression techniques have been known to produce such anomolous results in cases where the effect of one relevant parameter has been submerged in another.

This method is used in some mid-western states of the USA in which a considerable number of catchment areas are in the same climatic region. Results are not transferable to different climatic regions and the equation must be verified before use for predictive purposes.

The simplest general formula is one which relates flood flows to basin area. For example, in some of the Canadian prairie provinces, it is possible to relate the peak flood to the area and recurrence interval by

$$Q_{tr} = C(32.3\,A^{0.5}\,tr^{0.444}) \qquad (6.8)$$

where Q_{tr} = peak flood in cfs for return period tr years, A = drainage area (miles^2) and C = runoff coefficient dependent on watershed characteristics.

More complex equations may, however, be necessary and an equation for the daily mean flood peak in one part of eastern Canada relates the flood flow to recurrence interval, area of catchment, mean annual runoff (mm), the median snow on the ground at the end of March (mm) and the longitude in degrees.

6.5   The rational method

This is an approximate method of calculating surface runoff widely used in urban areas and small catchments (< 100 square miles). It is based on the principle that if there were no losses due to evaporation, transpiration, soil moisture, etc. the flow would be given by

$$Q = IA \qquad (6.9)$$

where I = intensity and A = area of catchment. Since some water will be lost, a runoff coefficient, C is introduced. Thus

$$Q = CIA \qquad (6.10)$$

In its simplest form where there is only one drainage channel the procedure is fairly straightforward. First an estimate is made of the time of concentration T_c. This is the time it takes a particle of water to travel from the most hydraulically remote point on the catchment

Table 6.1 Typical runoff coefficients

Description of area	Runoff coefficients
Downtown business areas	0.70–0.95
Residential (suburban)	0.25–0.40
Apartment dwelling areas	0.50–0.70
Light industrial areas	0.50–0.80
Heavy industrial areas	0.60–0.90
Parks, cemeteries	0.10–0.25
Unimproved areas	0.10–0.30
Streets	0.70–0.90
Roofs	0.75–0.95
Flat lawns; sandy soil	0.05–0.10
Flat lawns; heavy soil	0.13–0.17

to the outfall. It includes the time taken for water to travel overland into the drainage channel and also the time during which the water is in the channel. The duration of the design storm is chosen to be equal to the time of concentration and the intensity would be obtained from an intensity-duration-frequency curve such as that shown in Figure 5.1.

The value of the runoff coefficient depends on the conditions of the area which is under study. Table 6.1 provides typical values. Having determined the runoff coefficient and the storm intensity, peak discharge for the catchment may be evaluated from Equation (6.10).

WORKED EXAMPLES

Example 6.1   WEIRDIS: discharge over weirs

Write a program to permit calculation of flow over a variety of weirs based on a measured head. The program should be capable of dealing with V-notch weirs, rectangular weirs, with and without side contractions, and broad-crested weirs. In the case of the broad crested weir, assume no energy loss and still water upstream.

The program must be interactive and allow the user initially to choose which type of weir. Having made the choice, the user will then supply a value for the head over the weir in order to obtain the discharge. One way of doing this is to set up a number of subroutines, one routine for each weir.

```
10 CLS
20 PRINT "DISCHARGE OVER WEIRS"
30 PRINT "--------------------"
40 PRINT
50 PRINT
60 PRINT "OPTIONS : " : PRINT
70 PRINT 1,"V-NOTCH WEIRS"
80 PRINT 2,"RECTANGULAR WEIRS (No Side Restrictions)"
90 PRINT 3,"RECTANGULAR WEIRS (Side Restrictions)"
100 PRINT 4,"BROAD CRESTED WEIRS"
110 PRINT 5,"END APPLICATION"
120 PRINT : PRINT : INPUT "CHOICE"; C$ : CH = VAL(C$)
130 IF (CH < 1 OR CH > 5) THEN PRINT "BAD CHOICE NUMBER !" : FOR I = 1 TO 1000 : NEXT I : GOTO 10
140 ON CH GOSUB 190,360,550,710,150 : GOTO 10
150 PRINT "END...CONFIRM (Y/N)?";
160 INPUT A$ : IF A$ = "Y" THEN END
165 GOTO 10
170 :
180 :
190 REM               *** V-NOTCH WEIRS
200 CLS
210 PRINT "V-NOTCH WEIRS"
220 PRINT "-------------"
230 PRINT
240 INPUT "WHAT IS ANGLE OF NOTCH IN DECIMAL DEGREES "; A
250 PRINT
260 INPUT "HEIGHT OF WATER ABOVE NOTCH IN meters "; E
270 PRINT
280 INPUT "WHAT IS THE DISCHARGE COEFFICIENT ( e.g. 0.585 ) "; C
```

```
290 PRINT
300 R = A # .0174532930
310 V = (8/15) # C # SQR(219.806001) # TAN(R/2) # (E ^ (5/2) )
320 GOSUB 840
330 RETURN
340 :
350 :
360 REM                    ### RECTANGULAR WEIR (NO SIDE RESTRICTIONS)
370 CLS
380 PRINT "RECTANGULAR WEIR (No Side Restrictions)"
390 PRINT "----------------------------------------"
400 PRINT
410 INPUT "WHAT IS THE CHANNEL WIDTH IN meters "; W
420 PRINT
430 INPUT "WHAT IS THE HEIGHT OF WEIR IN meters "; H
440 PRINT
450 INPUT "WHAT IS THE UPSTREAM DEPTH IN meters "; D
460 PRINT
470 E = D-H
480 IF E < 0 THEN PRINT "NO FLOW ! CHECK INPUT AND TRY AGAIN !" : FOR I = 1 TO 1000 : NEXT I : GOTO 360
490 C = .611 + ( .08 # (E/H) )
500 V = W # (2/3) # C # SQR( 219.806001 ) # ( E ^ (3/2) )
510 GOSUB 840
520 RETURN
530 :
540 :
550 REM                    ### RECTANGULAR CHANNEL (Side Restrictions)
560 CLS
570 PRINT "RECTANGULAR WEIRS (Side Restrictions)"
580 PRINT "----------------------------------------"
590 PRINT
600 INPUT "WHAT IS THE WEIR OPENING IN meters "; L
610 PRINT
620 INPUT "WHAT IS THE HEIGHT OF WATER ABOVE THE WEIR IN meters "; H
630 PRINT
640 INPUT "WHAT IS THE COEFFICIENT OF DISCHARGE (i.e. 0.611) "; C
650 PRINT
660 V = (2/3) # C # ( L - ( .2 # H ) ) # SQR(2 # 9.806001) # ( H ^ (3/2) )
670 GOSUB 840
680 RETURN
690 :
700 :
710 REM               ### BROAD CRESTED WEIR
720 CLS
730 PRINT "BROAD CRESTED WEIR "
740 PRINT "------------------ "
750 PRINT
760 INPUT "WHAT IS THE CHANNEL WIDTH IN meters "; W
770 PRINT
780 INPUT "WHAT IS THE UPSTREAM HEIGHT ABOVE THE CREST IN meters "; H
790 V = (2/3) # H # SQR( (2/3) # 9.806001 # H ) # W
800 PRINT : PRINT
810 GOSUB 840
820 RETURN
830 :
840 REM               ### PRINT AND RETURN SUBROUTINE
850 PRINT "THE VOLUME OF FLOW IS "; V ;" cu meters / sec "
860 PRINT : PRINT : PRINT
870 PRINT "PRESS RETURN KEY < enter > TO CONTINUE ! ";
880 INPUT A$ : GOTO 10


RUN

DISCHARGE OVER WEIRS
--------------------


OPTIONS :

   1           V-NOTCH WEIRS                               4          BROAD CRESTED WEIRS
   2           RECTANGULAR WEIRS (No Side Restrictions)    5          END APPLICATION
   3           RECTANGULAR WEIRS (Side Restrictions)
```

```
CHOICE? 1

V-NOTCH WEIRS
-------------

WHAT IS ANGLE OF NOTCH IN DECIMAL DEGREES ? 90

HEIGHT OF WATER ABOVE NOTCH IN meters ? 1

WHAT IS THE DISCHARGE COEFFICIENT ( e.g. 0.585 ) ? .585

THE VOLUME OF FLOW IS  1.381706  cu meters / sec


PRESS RETURN KEY ( enter )  TO CONTINUE ! ?


DISCHARGE OVER WEIRS
--------------------

OPTIONS :

    1           V-NOTCH WEIRS
    2           RECTANGULAR WEIRS (No Side Restrictions)
    3           RECTANGULAR WEIRS (Side Restrictions)
    4           BROAD CRESTED WEIRS
    5           END APPLICATION

CHOICE? 3


RECTANGULAR WEIRS (Side Restrictions)
-------------------------------------

WHAT IS THE WEIR OPENING IN meters ? 4

WHAT IS THE HEIGHT OF WATER ABOVE THE WEIR IN meters ? 2.5

WHAT IS THE COEFFICIENT OF DISCHARGE (i.e. 0.611) ? .611

THE VOLUME OF FLOW IS  24.9568  cu meters / sec

PRESS RETURN KEY ( enter )  TO CONTINUE ! ?


DISCHARGE OVER WEIRS
--------------------

OPTIONS :

    1           V-NOTCH WEIRS
    2           RECTANGULAR WEIRS (No Side Restrictions)
    3           RECTANGULAR WEIRS (Side Restrictions)
    4           BROAD CRESTED WEIRS
    5           END APPLICATION

CHOICE? 5
END...CONFIRM (Y/N)?? Y
Ok
```

Program notes

(1) Lines 0-30 — print title.
(2) Lines 40-130 — menu define choice of five actions: four types of weirs and stop.
(3) Lines 140-180 — select choice and confirm end option.
(4) Lines 190-230 — print title (V-Notch weir)
(5) Lines 240-290 — enter variables for calculations.
(6) Lines 300-310 — calculate flow.
(7) Lines 320-350 — subroutine prints flow and returns to menu. (Line 840 for subroutine)
(8) Lines 360-400 — print title (rectangular weir — n.s.r.).
(9) Lines 410-460 — enter variables for calculation.
(10) Lines 470-480 — check for flow.
(11) Lines 490-500 — calculate flow.
(12) Lines 510-520 — subroutine prints flow and returns to menu. (Line 840 for subroutine).
(13) Lines 550-590 — print title (rectangular weir — s.r.)
(14) Lines 600-650 — enter variables for calculation.
(15) Line 660 — calculates flow.
(16) Lines 670-680 — Subroutine prints flow and returns to menu. (Line 840 for subroutine).
(17) Lines 710-750 — print title (broad crested weir)
(18) Lines 760-780 — enter variables for calculations.
(19) Lines 790-800 — calculate flow.
(20) Lines 810-820 — subroutine prints flow and returns to menu. (Line 840 for subroutine)
(21) Lines 840-880 — subroutine to print flow and returr to menu.

Example 6.2   STAGEDIS: calculation of discharge from velocity measurements

The data provided below were obtained from discharge measurements on a large river. The river was divided into a number of sections as shown in Figure 6.7 and the area of each section was measured. Table 6.2 below provides details of area and the results of a velocity traverse in each section. Develop a program to calculate the total discharge in the river.

The discharge in each section of the river can be calculated by multiplying the area of the section by the average velocity. The total discharge is given by the summation of the discharges in each section. The program should be developed to handle the general problem covering different numbers of sections and different numbers of velocity measurements in each section.

Table 6.2 Velocity traverse data

Section	1	2	3	4	5	6	7
Area (m)^2	3.1	5.1	7.1	9.2	6.6	4.8	2.6
Velocities (m/s)	0.65	0.7	0.7	0.75	0.65	0.6	0.6
	0.8	0.85	0.84	0.89	0.85	0.71	0.7
		0.84	0.82	0.87	0.85	0.70	0.62
			0.8	0.78	0.7		

```
10 CLS
20 PRINT "STAGE DISCHARGE - Calculation of discharge from velocity measurements "
30 PRINT "-----------------------------------------------------------------------"
40 PRINT : PRINT : PRINT
50 S = 0 : AV = 0
60 INPUT "ENTER NUMBER OF SECTIONS "; N
70 PRINT
80 INPUT "MAXIMUM NO. OF VELOCITY MEASUREMENTS IN ANY SECTION EQUALS "; T
90 DIM V(N),VL(T),AR(N)
100 FOR I = 1 TO N
110 CLS
120 PRINT "SECTION NUMBER "; I
130 PRINT
140 INPUT "ENTER AREA OF SECTION ( sq. meters ) "; A
150 AR(I) = A
160 PRINT
170 INPUT "ENTER NO. OF VELOCITY MEASUREMENTS IN THIS SECTION "; P
180 PRINT
190 FOR J = 1 TO P
200 PRINT "VELOCITY NO. "; J ;" = ";
210 INPUT VL(J)
220 NEXT J
230 PRINT
240 S = VL(1)
250 FOR J = 2 TO P
260 S = S + VL(J)
270 NEXT J
280 AV = S / P
290 V(I) = AV * A
300 NEXT I
310 CLS
320 PRINT "SECTION        AREA        DISCHARGE"
330 PRINT "-------        ----        ---------"
340 T1 = V(1)
350 FOR I = 1 TO (N-1)
360 PRINT I, AR(I), V(I)
370 T1 = T1 + V(I+1)
380 NEXT I
390 PRINT N, AR(N), V(N)
400 PRINT : PRINT : PRINT
410 PRINT "TOTAL DISCHARGE EQUALS "; T1 ;" cu. meters / sec "
420 PRINT : PRINT : PRINT
430 INPUT "DO YOU WISH TO DO THIS PROGRAM AGAIN (Y/N) "; A$
440 IF A$ = "Y" THEN CLEAR : GOTO 10
450 END
```

```
RUN

STAGE DISCHARGE - Calculation of discharge from velocity measurements
------------------------------------------------------------------------

ENTER NUMBER OF SECTIONS ? 7

MAXIMUM NO. OF VELOCITY MEASUREMENTS IN ANY SECTION EQUALS ? 4


SECTION NUMBER  1

ENTER AREA OF SECTION ( sq. meters ) ? 3.1

ENTER NO. OF VELOCITY MEASUREMENTS IN THIS SECTION ? 2

VELOCITY NO.  1  = ? .65
VELOCITY NO.  2  = ? .80



SECTION NUMBER  2

ENTER AREA OF SECTION ( sq. meters ) ? 5.1

ENTER NO. OF VELOCITY MEASUREMENTS IN THIS SECTION ? 3

VELOCITY NO.  1  = ? .70
VELOCITY NO.  2  = ? .85
VELOCITY NO.  3  = ? .84



SECTION NUMBER  3

ENTER AREA OF SECTION ( sq. meters ) ? 7.1

ENTER NO. OF VELOCITY MEASUREMENTS IN THIS SECTION ? 4

VELOCITY NO.  1  = ? .70
VELOCITY NO.  2  = ? .84
VELOCITY NO.  3  = ? .82
VELOCITY NO.  4  = ? .80



SECTION NUMBER  4

ENTER AREA OF SECTION ( sq. meters ) ? 9.2

ENTER NO. OF VELOCITY MEASUREMENTS IN THIS SECTION ? 4

VELOCITY NO.  1  = ? .75
VELOCITY NO.  2  = ? .89
VELOCITY NO.  3  = ? .87
VELOCITY NO.  4  = ? .78
```

```
SECTION NUMBER  5

ENTER AREA OF SECTION ( sq. meters ) ? 6.6

ENTER NO. OF VELOCITY MEASUREMENTS IN THIS SECTION ? 4

VELOCITY NO.  1  = ? .65
VELOCITY NO.  2  = ? .85
VELOCITY NO.  3  = ? .85
VELOCITY NO.  4  = ? .7



SECTION NUMBER  6

ENTER AREA OF SECTION ( sq. meters ) ? 4.8

ENTER NO. OF VELOCITY MEASUREMENTS IN THIS SECTION ? 3

VELOCITY NO.  1  = ? .60
VELOCITY NO.  2  = ? .71
VELOCITY NO.  3  = ? .70



SECTION NUMBER  7

ENTER AREA OF SECTION ( sq. meters ) ? 2.6

ENTER NO. OF VELOCITY MEASUREMENTS IN THIS SECTION ? 3

VELOCITY NO.  1  = ? .60
VELOCITY NO.  2  = ? .70
VELOCITY NO.  3  = ? .62



SECTION     AREA      DISCHARGE
-------     ----      ---------
   1         3.1       2.2475
   2         5.1       4.063
   3         7.1       5.609
   4         9.2       7.567
   5         6.6       5.0325
   6         4.8       3.216
   7         2.6       1.664

TOTAL DISCHARGE EQUALS  29.399  cu. meters / sec

DO YOU WISH TO DO THIS PROGRAM AGAIN (Y/N) ? N
```

Program notes

(1) Lines 0-40 — print title.
(2) Lines 50-90 — set constants, enter maximum number of observations and dimension arrays.

(3) Lines 100-300 — data input and calculation section.
(4) Lines 110-180 — for each section; enter area and number of velocity measurements.
(5) Lines 190-230 — enter each velocity measurement.
(6) Lines 240-290 — calculate average velocity and discharge for each section.
(7) Lines 310-330 — print headings.
(8) Lines 340-420 — calculate total discharge and print by section and total.
(9) Lines 430-450 — permit recalculation with new data.

Example 6.3 EFFRAIN: calculation of effective rainfall from hydrograph

Figure 6.8 is a hydrograph showing the flow in a river following a storm which covered the entire catchment. The catchment area was 1250 km². Calculate the depth of effective rainfall.

The hydrograph may be considered to be a composite of surface

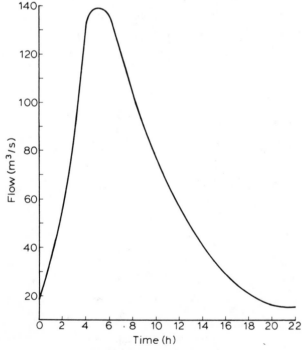

Figure 6.8 Inflow hydrograph

runoff and groundwater flow. When these are separated the area under the surface runoff hydrograph provides a measure of the volume of surface runoff. This volume is equal to the product of catchment area and depth of effective rainfall. The program must permit the user to deduct groundwater flow from the ordinates of the given hydrograph. The area of the surface runoff hydrograph must then be measured using some numerical technique and must be divided by the catchment area to give the depth of effective rainfall.

```
10 REM CALCULATION OF EFFECTIVE RAINFALL FROM HYDROGRAPH
20 CLS
30 PRINT "EFFECTIVE RAINFALL FROM HYDROGRAPH"
35 PRINT "---------------------------------"
40 PRINT
50 INPUT "NUMBER OF OBSERVATIONS "; N
60 DIM A1(N),A2(N),A3(2*N)
70 PRINT "INSTRUCTIONS : ENTER FLOWS (in m^3/s) AT EVEN INTERVALS "
80 INPUT "INTERVAL IN HOURS "; HR
90 FOR I = 0 TO N-1
100 PRINT "AT TIME "; I * HR ;" FLOW EQUALS ";
110 INPUT A1(I)
120 NEXT I
130 PRINT
140 PRINT "IS BASE FLOW CONSTANT ( Y/N ) ";
144 INPUT A$
146 IF A$ = "N" THEN GOSUB 2000 : GOTO 230
150 INPUT "ENTER CONSTANT BASE FLOW "; CB
160 K = 0
170 FOR I = 0 TO N-1
180 T = A1(I) - CB
190 IF T < 0 THEN GOTO 220
200 A2(K) = T
210 K = K + 1
220 NEXT I
230 GOSUB 1000
250 INPUT "CATCHMENT AREA (sq. meters) "; AR
260 DP# = VL# / AR
270 DC = DP# * 100
280 REM                    *** OUTPUT RESULTS
290 CLS
300 PRINT "EFFECTIVE RAINFALL FROM HYDROGRAPH"
310 PRINT "---------------------------------"
320 PRINT
330 PRINT "VOLUME OF RAINFALL  ( m^3 ) "; VL#
340 PRINT "CATCHMENT AREA  ( m^2 ) "; AR
350 PRINT : PRINT : PRINT
360 PRINT "EFFECTIVE RAINFALL IN cm IS "; DC
370 PRINT : PRINT : PRINT
380 PRINT "DO YOU WISH TO DO THIS PROGRAM AGAIN ";
390 INPUT A$ : IF A$ = "Y" THEN CLEAR : GOTO 10
400 END
1000 REM SIMPSON'S RULE
1010 T = 0
1020 C = 1
```

```
1030 X1 = A2(0)
1040 X2 = A2(K)
1050 FOR I = 1 TO (2*K)-1
1060 TP = I MOD 2
1070 IF TP <> 0 THEN GOTO 1100
1080 A3(I) = A2(I\2)
1090 GOTO 1120
1100 A3(I) = ( A2(C) + A2( CINT(C-1) )  ) / 2
1110 C = C + 1
1120 NEXT I
1130 FOR I = 1 TO (2*K)-1
1140 TP = I MOD 2
1150 IF TP <> 0 THEN GOTO 1180
1160 T = T + ( 2 * A3(I) )
1170 GOTO 1190
1180 T = T + ( 4 * A3(I) )
1190 NEXT I
1200 S = (( HR / 2) / 3 ) * ( X1 + T + X2 )
1210 TS# = ( K-1) * HR * ( 60^2 )
1220 VL# = S * TS#
1230 RETURN
2000 REM NON-CONSTANT BASE FLOW
2010 K = 0
2020 FOR I = 0 TO N-1
2030 PRINT "AT TIME "; I * HR ;" TOTAL FLOW IS "; A1(I) ;" BASE FLOW EQUALS ";
2040 INPUT B
2050 T = A1(I) - B
2060 IF T < 0 THEN GOTO 2090
2070 A2(K) = T
2080 K = K + 1
2090 NEXT I
2100 RETURN
```

```
RUN


EFFECTIVE RAINFALL FROM HYDROGRAPH
----------------------------------


NUMBER OF OBSERVATIONS ? 12
INSTRUCTIONS : ENTER FLOWS (in m^3/s) AT EVEN INTERVALS
INTERVAL IN HOURS ? 2
AT TIME   0  FLOW EQUALS ? 10
AT TIME   2  FLOW EQUALS ? 57
AT TIME   4  FLOW EQUALS ? 133
AT TIME   6  FLOW EQUALS ? 136
AT TIME   8  FLOW EQUALS ? 102
AT TIME  10  FLOW EQUALS ? 76
AT TIME  12  FLOW EQUALS ? 56
AT TIME  14  FLOW EQUALS ? 41
AT TIME  16  FLOW EQUALS ? 28
AT TIME  18  FLOW EQUALS ? 18
AT TIME  20  FLOW EQUALS ? 12
AT TIME  22  FLOW EQUALS ? 10
```

```
IS BASE FLOW CONSTANT ( Y/N )  ? Y
ENTER CONSTANT BASE FLOW ? 10
CATCHMENT AREA (sq. meters) ? 1250 E+06



EFFECTIVE RAINFALL FROM HYDROGRAPH
----------------------------------

VOLUME OF RAINFALL  ( m^3 )  88545600
CATCHMENT AREA  ( m^2 )  1.25E+09



EFFECTIVE RAINFALL IN cm IS  7.083648



DO YOU WISH TO DO THIS PROGRAM AGAIN ? N
```

Program notes

(1) Lines 0-40 — clear screen and print title.
(2) Lines 50-60 — input number of observations and dimension arrays.
(3) Lines 70-80 — instruction and input of interval.
(4) Lines 90-120 — loop to input flows.
(5) Lines 130-160 — determine base flow (subroutine used if not constant).
(6) Lines 170-220 — calculations with constant base flow.
(7) Line 230 — GOSUB 1000 — subroutine uses Simpson's rule to evaluate area (volume) under curve.
(8) Lines 250-270 — input catchment area and determine depth of rain.
(9) Lines 280-370 — print out results.
(10) Lines 380-400 — check to see if repeat calculation is required.
(11) Lines 1000-1230 — Simpson's rule subroutine.
(12) Lines 2000-2100 — routine to handle varying base flow.

Example 6.4 SRUN: calculation of surface runoff

Equation (6.11) below relates the daily mean flood peak to a number of catchment variables in catchments in eastern Canada.

$$Q_D = K + a \ln A + b \ln MAR + c \ln MSM \\ + d \ln \text{Long} + e \ln DF \tag{6.11}$$

where Q_d = daily mean flood peak, A = area (km^2), DF = drainage factor (number of confluences/km^2), MAR = mean annual runoff (mm), *MSM* = median snow on ground at end of March (mm) and Long = longitude (degrees). Values of the constants in this equation vary with recurrence interval and are shown below in Table 6.3. Develop a program which will permit evaluation of the daily mean flood peak for different recurrence intervals and different values of the relevant variables.

Table 6.3 Values of constants in Equation (6.11)

Recurrence interval (years)	Constants					
	K	a	b	c	d	e
2	65.16	0.91	−1.10	0	−14.03	0.19
20	44.99	0.84	0	0.15	−10.92	0.14
100	43.36	0.83	0	0.18	−10.48	0.12

An interactive program will be developed permitting the user to choose the recurrence interval and to set the values of the variables (i.e. area, longitude, etc.). Values of the constants will be specified on the basis of the recurrence interval and will be as shown in Table 6.3.

```
10 CLS
20 PRINT "SURFACE RUNOFF "
30 PRINT "-------------- "
40 PRINT
50 INPUT "ENTER RECURRENCE INTERVAL ( 2, 20 OR 100 years ) "; N
60 IF N = 2 THEN GOTO 290
70 IF N = 20 THEN GOTO 310
80 IF N = 100 THEN GOTO 330
90 PRINT "INCORRECT RECURRENCE INTERVAL - TRY AGAIN ! "
100 FOR I = 1 TO 500 : NEXT I
110 GOTO 10
120 PRINT
130 INPUT "ENTER AREA OF CATCHMENT ( sq. km ) "; AR
140 PRINT
150 INPUT "ENTER DRAINAGE FACTOR ( no. of confluences / sq. km ) "; DF
160 PRINT
170 INPUT "ENTER MEAN ANNUAL RUNOFF ( mm ) "; MR
180 PRINT
190 INPUT "ENTER MEDIAN SNOW ON GROUND AT END OF MARCH ( mm ) "; MS
200 PRINT
210 INPUT "ENTER LONGITUDE ( decimal degrees ) "; L
220 PRINT : PRINT : PRINT
230 F = K + ( A * LOG(AR) ) + ( B * LOG(MR) ) + ( C * LOG(MS) ) + ( D * LOG(L) ) + ( E * LOG(DF) )
240 PRINT "THE DAILY MEAN FLOOD PEAK IS "; F ;" cu. meters / sec "
250 PRINT : PRINT : PRINT
260 INPUT "DO YOU WISH TO DO THIS PROGRAM AGAIN (Y/N) "; A$
```

```
270 IF A$ = "Y" GOTO 10
280 END
290 K = 65.16 : A = .9099999 : B = -1.1 :C = 0 : D = -14.03 : E = .19
300 GOTO 120
310 K = 44.99 : A = .84 : B = 0 :C = .15 : D = -10.92 : E = .14
320 GOTO 120
330 K = 43.36 : A = .83 : B = 0 :C = .18 : D = -10.48 : E = .12
340 GOTO 120


RUN


SURFACE RUNOFF
--------------

ENTER RECURRENCE INTERVAL ( 2, 20 OR 100 years ) ? 20

ENTER AREA OF CATCHMENT ( sq. km ) ? 54

ENTER DRAINAGE FACTOR ( no. of confluences / sq. km ) ? 2

ENTER MEAN ANNUAL RUNOFF ( mm ) ? 870

ENTER MEDIAN SNOW ON GROUND AT END OF MARCH ( mm ) ? 965

ENTER LONGITUDE ( decimal degrees ) ? 66


THE DAILY MEAN FLOOD PEAK IS  3.717574  cu. meters / sec


DO YOU WISH TO DO THIS PROGRAM AGAIN (Y/N) ? N
```

Program notes

(1) Lines 0-40 — print title.
(2) Lines 50-120 — select recurrence interval, input constants and check for bad interval.
(3) Lines 130-220 — enter variables for calculation.
(4) Lines 230-240 — calculate and print magnitude of flood.
(5) Lines 250-280 — permit recalculation with new data.
(6) Lines 290-340 — constants used in calculations based on recurrence interval.

PROBLEMS

(6.1) Worked example 6.1 assumes still water upstream of a broad-crested weir with no energy loss. Modify the program to account for a variable loss and to include the upstream velocity head.

(6.2) Example 6.4 calculates flood flows on the basis of an empirical equation. Develop a similar program based on a formula suitable for catchments in your local area and compare results with data published for your area.

(6.3) In small streams and fast-flowing streams in mountain areas, it is often impossible to measure streamflow by current meter gauging. Under such circumstances chemical gauging, using some common chemical as a tracer may be advantageous. If a tracer of concentration C_1 is injected into the stream at a rate, q, then after full mixing the equilibrium concentration, C_e, is related to the flow in the stream Q and the other variables, by

$$C_e(Q+q) = C_1 q$$

Use this equation to develop a program suitable for use in a chemical gauging process.

(6.4) Given below are daily mean flows over a five-day period.

Day	1	2	3	4	5
Flow	300	2500	1600	1000	650

Develop a program to calculate the mean flow rate, the total discharge volume and the depth of runoff over a catchment area of 350 square units. The program should be generalised to deal with up to 10 days and to handle input in the form of ft³/s or m³/s for discharge and sq. miles or sq. km for area. The units of total discharge should be cusec-days or Acre ft (chosen by operator) for flow specified as ft³/s and m³ for flow specified as m³/s. Depth of runoff should be specified as inches or mm.

(6.5) An urban area is approximately rectangular (10 000 ft × 5000 ft) with a main drain running diagonally across the area. The difference in elevation from one end of the drain to the other is 12.5 ft.

The time of concentration for the area may be estimated from

$$Tc_{min} = \frac{0.0078 \, L^{0.77}}{S^{0.385}}$$

where L = max length of travel (ft) and S = average slope, highest-lowest point. Runoff coefficients vary throughout the area so that 40% of area is at 0.4; 40% of area is at 0.65; and 20% of area is at 0.8.

Locate an intensity duration frequency diagram for your locality and use it to determine the peak rate of runoff from the above area for a storm having a return period of 25 years.

Develop a program to assist your calculations and to examine the effects of varying size of area, elevation of drain and storm return period.

Chapter 7

The unit hydrograph

ESSENTIAL THEORY

7.1 Introduction

The unit hydrograph for a catchment area is a typical hydrograph for that particular area. It correlates rainfall with runoff and indicates the runoff which would be experienced from one unit of rain falling evenly over the catchment. The correlation is made between effective or net rain (i.e. that portion of rain remaining after all losses by evaporation, interception and infiltration, etc. have been allowed for) and surface runoff (i.e. the hydrograph of runoff after the base flow has been abstracted).

The method is generally used in catchments varying in area from 100 - 2000 square miles. Good flood and precipitation records are required to clearly indicate which storm caused which flood.

7.2 Derivation of unit graph

The derivation of the unit hydrograph is based primarily on two assumptions

(1) That the duration of runoff from a uniform intensity storm depends only on the duration of the storm and is independent of the intensity of the rain falling. In practice, this assumption has been found to be generally true.

(2) That the ordinates of the surface runoff hydrograph resulting from a uniform intensity storm are directly proportional to the intensity of that storm. For example, if the intensity of the rainfall is doubled then the magnitude of the hydrograph ordinates will also be doubled.

The derivation of the unit hydrograph requires an examination of the rainfall and runoff record in order to isolate single runoff events caused by individual, isolated, storms. Ideally the storm should have a fairly uniform intensity throughout the period of rainfall excess and should be distributed fairly evenly over the entire catchment. If possible, a number of such storms should be analysed. The unit graph is developed from the runoff hydrographs in the following manner.

First the base flow is separated to give the hydrograph of surface runoff (as described in Chapter 6). Measurement of the area under the curve of surface runoff provides a measure of the volume of surface runoff. Because this originated from the rainfall excess, the depth of effective rain may be calculated by dividing the volume of surface runoff by the area of the catchment. The ordinates of the runoff hydrograph are assumed to be directly proportional to the depth of effective rain. Thus, dividing the measured ordinates by the calculated depths gives the ordinates which would result from one unit of rain. This unit may be specified as 1in, 1cm or 1mm. When this has been done, the resulting hydrograph is the unit hydrograph for a duration equal to that of the original storm. This process is illustrated diagramatically in Figure 7.1.

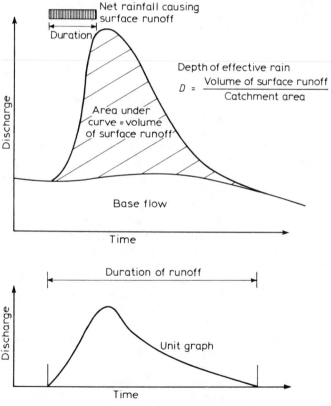

Figure 7.1   Derivation of unit hydrograph

In practice it will be found that the storms used for deriving the unit hydrograph will be of different durations so that a number of unit hydrographs, each of a specified but different duration will be obtained. To obtain one, standard, unit hydrograph for the basin, it will be necessary to convert the duration of each of these hydrographs to one standard — say one hour, four fours, etc. The final unit hydrograph may then be obtained by averaging the ordinates of each unit hydrograph. Some adjustment of the result to obtain one unit of direct runoff may be necessary.

The use of the unit hydrograph in developing a design flood from a design storm effectively reverses the process. This will be described later.

7.3   Changing the duration to a multiple of the original

Modification of the duration of a given unit hydrograph to an integer multiple (e.g. changing a two-hour duration to four, six, eight hours, etc.) requires the use of a third basic assumption in the unit graph process.

(3) Hydrographs resulting from different storms occurring in sequence, with or without periods of zero rainfall, may be superimposed to create a composite hydrograph of the total storm.

Suppose, for example, that a unit hydrograph is available for a four-hour duration storm and that it is required to convert this to an eight-hour duration storm. The four-hour unit graph results from one unit of rain falling over a four-hour period and would plot with the flow starting at time zero and finishing at some time which would depend on the characteristics of the catchment. If another unit storm, again giving one unit of rain over a four-hour period, started at hour four, it would be possible to plot a second unit hydrograph identical to the first but with the flow starting at hour four rather than hour zero. These two could then be superimposed and added, as shown in Figure 7.2, to give the hydrograph resulting from one unit of rain in the first four hours and one unit of rain in the second four hours. This, then, would be the hydrograph resulting from a uniform intensity storm which delivered two units of rain over an eight-hour period. By dividing the ordinates of the resulting hydrograph by two, a unit hydrograph for one unit of rain over an eight-hour period, i.e. an eight-hour unit hydrograph, will be obtained.

The process may be extended indefinitely. A twelve-hour hydrograph would be obtained by summating the ordinates of three unit graphs — one starting at hour zero, one starting at hour four

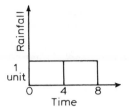

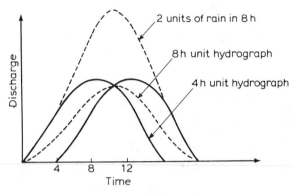

Figure 7.2   Multiple duration

and one starting at hour eight and then dividing the resulting, summated ordinates by three. Obviously, the method is not suitable for modifying the duration to any fraction or non-integer multiples of the original duration and other methods must be used in these cases. The process of straightforward superposition may also be cumbersome if a conversion is required from a short duration to a very long duration. However, such a possibility is rare and relatively unlikely to be of much practical interest.

7.4   Conversion to durations which are non-integer multiples or fractions of the original duration

In order to modify the duration of a unit hydrograph to a non-integer multiple, or to a shorter duration than that of the original, it is necessary to use an 'S' curve. This is defined as a hydrograph obtained for continuous rainfall at constant intensity. By successively lagging and summating all ordinates of the hydrographs (see Figure 7.3), the continuous flow hydrograph may be obtained. This has been termed an 'S' curve because of its shape. The process is illustrated numerically in Table 7.1.

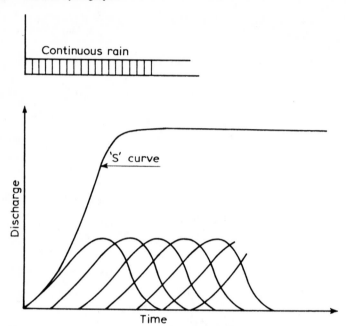

Figure 7.3   Development of 'S' curve

The first two columns give the ordinates of the unit graph having a duration equal to four hours. Columns three to eight show the lagged values, each column providing the ordinates for one unit of rain which started immediately the preceding rain ceased. The addition of all these ordinates (column ten) thus provides a measure of the hydrograph which would result from continuous, uniform intensity, rain. Because the unit graph had a four-hour duration, the intensity of the continuous storm is equivalent to 1in delivered in four hours, i.e. $\frac{1}{4}$ in/h. If the 'S' curve is now lagged by, for example, three hours the difference between the ordinates of the two 'S' curves will indicate the fall which would result from a storm having a duration of three hours at an intensity of $\frac{1}{4}$ in/h. This is shown in Figure 7.4 and may be verified by considering specific ordinates shown on that figure.

For example, at hour three the ordinate AB represents the fall from a storm having $\frac{1}{4}$ in/h for three hours. At hour eight the ordinate CD represents the flow for a continuous fall of $\frac{1}{4}$ in/h after eight hours but, ED represents the flow for a fall of $\frac{1}{4}$ in/h after five hours. The difference, ordinate CE, must then be the flow which would result from a fall of $\frac{1}{4}$ in/h for a duration of (8-5) three hours.

Table 7.1 Development of an 'S' curve

Time (h)	Discharge (m³/s)	'S' curve additions							'S' curves ordinates
0	0								0
2	47								47
4	123	0							123
6	126	47							173
8	92	123	0						215
10	66	126	47						239
12	46	92	123	0					261
14	31	66	126	47					270
16	18	46	92	123	0				279
18	8	31	66	126	47				278
20	2	18	46	92	123	0			281
22	0	8	31	66	126	47			278
24	0	2	18	46	92	123	0		281
		0	8	31	66	126	47		278
		0	2	18	46	92	123	0	281
			0	8	31	66	126		

The ordinal differences between the 'S' curves thus give the hydrograph resulting from $\frac{1}{4}$ in/h over three hours. This represents a total depth over the catchment of $\frac{3}{4}$ in over the period of the storm, i.e. three hours. To obtain the unit hydrograph for 1in in three hours, it is necessary to multiply the ordinal differences obtained from Figure 7.4 by $\frac{4}{3}$, i.e. if CE gives the flow for $\frac{3}{4}$ in over 3 h then CE $\times \frac{4}{3}$ is equal to the flow from 1in in three hours.

In general, to obtain a t hour duration unit graph from a D hour duration unit graph

(1)   Plot or tabulate the 'S' curve for the D hour unit graph.
(2)   Lag the 'S' curve by t hours and measure the ordinal differences.
(3)   Multiply these differences by D/t to get the ordinates of the t hour graph.

Here the 'S' Curve technique has been used to show how the duration may be changed to a fraction of the original. However, the same process is valid for any multiple. In Figure 7.4(b) the 'S' Curve

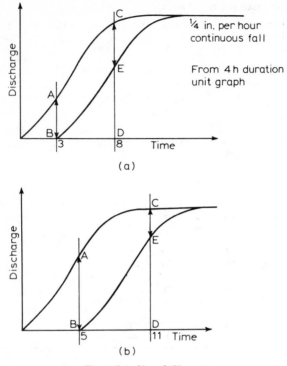

Figure 7.4 Use of 'S' curves

has been obtained from a two-hour duration unit graph and so shows the hydrograph which would result from a continuous storm having a uniform intensity of $\frac{1}{2}$ in/h. If this is lagged by five hours in order to obtain a five-hour unit hydrograph, then the ordinate CD (at hour eleven) represents the flow from a continuous storm of $\frac{1}{2}$ in/h over eleven hours. The ordinate ED represents the flow from $\frac{1}{2}$ in/h over six hours and the difference, CE, gives the flow for $\frac{1}{2}$ in/h over five hours. This represents a total rainfall of $2\frac{1}{2}$ in over the five-hour duration and the unit graph is obtained by multiplying the ordinate CE by $\frac{2}{5}$.

7.5   Application of unit graph

The use of a unit hydrograph to develop a synthetic flood hydrograph from a given design storm is based on the principle of superposition. First it is necessary to deduct all losses from the design storm in order to obtain the excess rainfall (it will be remembered that the unit hydrograph provides a correlation between surface runoff and net, effective, or excess rainfall). The duration of the unit hydrograph to be used must be equal to the duration of the design storm and this may necessitate the use of lagging or 'S' curves in order to obtain a suitable duration (it will be remembered that the duration of the unit hydrograph refers to the duration of the storm giving rise to the flood rather than to the period over which surface runoff occurs). If the storm is of uniform intensity it is then sufficient to multiply the ordinates of the appropriate unit hydrograph by the excess rainfall depths and to add in the base flow in order to obtain the full design hydrograph. However, in many cases, the design storm will be a composite, having one intensity in the first time period, a different intensity in the second and third time periods and so on. In that case the unit hydrograph may be used to obtain a surface runoff hydrograph for each separate component of the design storm, and these may then be summated as shown in Figure 7.5, to obtain the composite surface

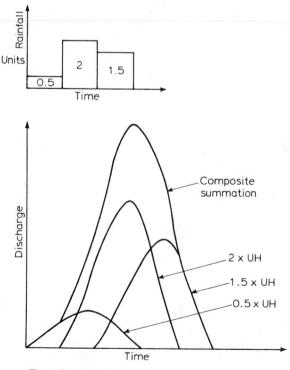

Figure 7.5   Development of composite hydrograph

runoff hydrograph. At that point the base flow may be added to provide the complete hydrograph.

WORKED EXAMPLES

Example 7.1 UGRAPH: derivation of unit hydrograph

The data given below provide details of the flood hydrograph recorded from a four-hour duration individual, isolated storm of fairly uniform intensity which was distributed evenly over the catchment. Write a program to calculate the ordinates of the unit hydrograph. (Area = 1250 km^2.)

Time (h)	0	2	4	6	8	10	12	14	16	18	20	22
Flow (m^3/s)	10	57	133	136	102	76	56	41	28	18	12	10

The program must read and store the given data. It must be possible to subtract the base flow in order to obtain the ordinates of the surface runoff hydrograph. The area under the curve must then be calculated using some numerical technique and this will give the volume of surface runoff. Dividing by the area of the catchment then gives the depth of effective rain. Dividing the ordinates of the surface runoff hydrograph by this depth then gives the unit graph.

```
10 REM UNIT HYDROGRAPH
20 CLS
30 PRINT "UNIT HYDROGRAPH"
40 PRINT "---------------"
50 PRINT
60 INPUT "NUMBER OF OBSERVATIONS "; N
70 DIM A1(N+1),A2(N+1),A3(2*N+1)
80 PRINT "INSTRUCTIONS : ENTER FLOWS (in m^3/s) AT EVEN INTERVALS "
90 INPUT "INTERVAL IN HOURS "; HR
100 FOR I = 0 TO N-1
110 PRINT "AT TIME "; I * HR :" FLOW EQUALS ";
120 INPUT A1(I)
130 NEXT I
140 PRINT
150 PRINT "IS BASE FLOW CONSTANT ( Y/N ) ";
152 INPUT A$
154 IF A$ = "N" THEN GOSUB 2000 : GOTO 250
160 INPUT "ENTER CONSTANT BASE FLOW "; CB
170 K = 0
180 FOR I = 0 TO N-1
190 T = A1(I) - CB
200 IF T < 0 THEN GOTO 230
210 A2(K) = T
220 K = K + 1
```

```
230 NEXT I
250 GOSUB 1000
260 INPUT "CATCHMENT AREA (sq. meters) "; AR
270 DP# = VL# / AR
280 DC = DP# * 100
290 FOR I = 0 TO K
300 A1(I) = A2(I) / DC
310 NEXT I
320 REM  OUTPUT RESULTS
330 CLS
340 PRINT "UNIT HYDROGRAPH"
350 PRINT "--------------"
360 PRINT
370 PRINT "TIME";TAB(15);"UNIT HYDROGRAPH ORDINATE"
380 PRINT "----";TAB(15);"-----------------------"
390 PRINT
400 FOR I = 0 TO K
410 PRINT I * HR ; TAB(20); A1(I)
420 NEXT I
430 PRINT : PRINT : PRINT
440 PRINT "DO YOU WISH TO DO THIS PROGRAM AGAIN ";
450 INPUT A$ : IF A$ = "Y" THEN CLEAR : GOTO 10
460 END
1000 REM SIMPSON'S RULE
1010 T = 0
1020 C = 1
1030 X1 = A2(0)
1040 X2 = A2(K)
1050 FOR I = 1 TO (2*K)-1
1060 TP = I MOD 2
1070 IF TP <> 0 THEN GOTO 1100
1080 A3(I) = A2(I\2)
1090 GOTO 1120
1100 A3(I) = ( A2(C) + A2( CINT(C-1) )  ) / 2
1110 C = C + 1
1120 NEXT I
1130 FOR I = 1 TO (2*K)-1
1140 TP = I MOD 2
1150 IF TP <> 0 THEN GOTO 1180
1160 T = T + ( 2 * A3(I) )
1170 GOTO 1190
1180 T = T + ( 4 * A3(I) )
1190 NEXT I
1200 S = (( HR / 2) / 3 ) * ( X1 + T + X2 )
1210 TS# = ( K-1) * HR * ( 60^2 )
1220 VL# = S * TS#
1230 RETURN
2000 REM NON-CONSTANT BASE FLOW
2010 K = 0
2020 FOR I = 0 TO N-1
2030 PRINT "AT TIME "; I * HR ;" TOTAL FLOW IS "; A1(I) ;" BASE FLOW EQUALS ";
2040 INPUT B
2050 T = A1(I) - B
2060 IF T < 0 THEN GOTO 2090
2070 A2(K) = T
2080 K = K + 1
2090 NEXT I
2100 RETURN
```

```
RUN

UNIT HYDROGRAPH
---------------

NUMBER OF OBSERVATIONS ? 12
INSTRUCTIONS : ENTER FLOWS (in m^3/s) AT EVEN INTERVALS
INTERVAL IN HOURS ? 2
AT TIME    0  FLOW EQUALS ? 10
AT TIME    2  FLOW EQUALS ? 57
AT TIME    4  FLOW EQUALS ? 133
AT TIME    6  FLOW EQUALS ? 136
AT TIME    8  FLOW EQUALS ? 102
AT TIME   10  FLOW EQUALS ? 76
AT TIME   12  FLOW EQUALS ? 56
AT TIME   14  FLOW EQUALS ? 41
AT TIME   16  FLOW EQUALS ? 28
AT TIME   18  FLOW EQUALS ? 18
AT TIME   20  FLOW EQUALS ? 12
AT TIME   22  FLOW EQUALS ? 10

IS BASE FLOW CONSTANT ( Y/N )  ? Y
ENTER CONSTANT BASE FLOW ? 10
CATCHMENT AREA (sq. meters) ? 1250 E+06

UNIT HYDROGRAPH
---------------

TIME       UNIT HYDROGRAPH ORDINATE
----       ------------------------

0          0
2          6.634999
4          17.36393
6          17.78745
8          12.98766
10         9.317233
12         6.493829
14         4.376276
16         2.541064
18         1.129362
20         .2823404
22         0
24         0

DO YOU WISH TO DO THIS PROGRAM AGAIN ? N
```

Program notes

(1) Lines 0-50 — clear screen and print title.
(2) Lines 60-70 — input number of observations and dimension arrays.
(3) Lines 80-90 — instruction and input of interval.
(4) Lines 100-140 — loop in input flow.

(5) Lines 150-160 — question for constant base flow. If yes — enter number. If no — GOSUB 2000. Subroutine non-constant base flow.
(6) Lines 170-230 — calculations with constant base flow.
(7) Line 250 — GOSUB 1000 — subroutine Simpson's rule evaluates area (volume) under curve.
(8) Lines 260-280 — input catchment area and determine depth of rain.
(9) Lines 290-310 — get ordinates of unit hydrograph by dividing by depth of rain.
(10) Lines 320-430 — output results.
(11) Lines 1000-1230 — subroutine Simpson's rule.
(12) Lines 2000-2100 — enter non-constant base flow and adjust hydrograph accordingly.

Note: This program is very similar to EFFRAIN. The depth of effective rainfall is calculated as before and this depth is used to modify the hydrograph of surface runoff.

Example 7.2   SCURVE: development of an 'S' curve

The unit hydrograph shown below resulted from a 4 h duration storm. Write a program to determine the ordinates of the associated 'S' curve.

T	0	4	8	12	16	20	24	28
Q	0	20	50	70	65	60	40	0

The process of developing an 'S' curve consists of successively lagging the ordinates of the unit graph by a time period equal to the unit graph duration and adding all ordinates to give the 'S' curve. This is shown in Table 7.1. Examination of that table shows that if only multiples of the original duration are considered (i.e. 0,4,8,12,16, etc.) then the ordintes of the 'S' curve may be obtained quickly by adding the hydrograph ordinate at any particular time to the sum of the ordinates at all previous times, i.e. the ordinate of the 'S' curve at time 16 h is given by adding that hydrograph ordinate to the sum of those at 12, 8, 4, and 0 h. This provides a quick and simple means of obtaining the 'S' curve in cases such as that given in the above example where the time period between ordinates is equal to the duration of the unit graph.

```
100 REM  S  CURVE DEVELOPMENT
200 CLS
210 PRINT "DEVELOPMENT OF S CURVE"
212 PRINT : PRINT : PRINT
```

```
225 PRINT "NUMBER OF ORDINATES = ";
230 INPUT N
231 DIM Q(N),S(N),T(N)
232 PRINT
235 FOR I = 1 TO N
240 PRINT " TIME "; I ;" = ";
245 INPUT T(I)
250 PRINT " DISCHARGE "; I ;" = ";
255 INPUT Q(I)
257 PRINT
260 NEXT I
265 S(1) = Q(1)
270 FOR I = 2 TO N
275 S(I) = S( I - 1 ) + Q(I)
280 NEXT I
285 PRINT " TIME        DISCHARGE      S VALUE"
287 PRINT" ------      -----------    ---------"
290 FOR I = 1 TO N
300 PRINT T(I),Q(I),S(I)
302 PRINT
305 NEXT I
310 END
```

RUN

DEVELOPMENT OF S CURVE

NUMBER OF ORDINATES = ? 8

```
 TIME  1  = ? 0
 DISCHARGE   1  = ? 0

 TIME  2  = ? 4
 DISCHARGE   2  = ? 20

 TIME  3  = ? 8
 DISCHARGE   3  = ? 50

 TIME  4  = ? 12
 DISCHARGE   4  = ? 70

 TIME  5  = ? 16
 DISCHARGE   5  = ? 65

 TIME  6  = ? 20
 DISCHARGE   6  = ? 60

 TIME  7  = ? 24
 DISCHARGE   7  = ? 40

 TIME  8  = ? 28
 DISCHARGE   8  = ? 0
```

TIME	DISCHARGE	S VALUE
0	0	0
4	20	20
8	50	70
12	70	140
16	65	205
20	60	265
24	40	305
28	0	305

Program notes

(1) Lines 0-210 — clear screen and print titles.
(2) Lines 225-232 — enter number of unit graph ordinates.
(3) Lines 235-260 — enter ordinate details.
(4) Lines 265-280 — calculate ordinates of 'S' curve.
(5) Lines 285-287 — print table titles.
(6) Lines 290-305 — print unit graph and 'S' curve ordinates.
(7) Line 310 — halt execution.

Example 7.3 MODUR: Modification of the duration using an 'S' curve

Using the data in Example 7.2 develop a program, using the 'S' curve technique, to obtain the unit hydrograph for a 12 h duration.

Alteration of the duration to a multiple may be undertaken by lagging the hydrograph as indicated in Section 7.3. However the methods discussed in Section 7.4 are valid although the method is usually used to handle fraction or non-integer multiples. To use the 'S' curve technique it it necessary to start the problem by developing an 'S' curve as was done in the previous example. The program developed to handle this aspect of the analysis may again be used. Modification is carried out by lagging the 'S' curve, finding the ordinal differences and multiplying by the ratio of the original duration/new duration. The program developed in Example 7.2 must therefore be extended to handle these calculations.

```
100 REM  S CURVE DEVELOPMENT - UNIT HYDROGRAPH
200 CLS
210 PRINT "DEVELOPMENT OF S CURVE"
212 PRINT : PRINT : PRINT
215 PRINT "NUMBER OF ORDINATES = ";
217 INPUT N
219 PRINT : PRINT
220 DIM DS(N + 10), S(N + 10),T(N+10), Q(N+10)
225 PRINT "TIME INCREMENT = ";
227 INPUT DT
230 PRINT : PRINT
235 FOR I = 1 TO N
240 PRINT "TIME "; I ;" =";
245 INPUT T(I)
250 PRINT "DISCHARGE "; I ;" =";
255 INPUT Q(I)
257 PRINT
260 NEXT I
265 S(1) = Q(1)
270 FOR I = 2 TO N
275 S(I) = S(I-1) + Q(I)
```

```
280 NEXT I
285 PRINT " TIME     DISCHARGE     S VALUE"
287 PRINT "------   ----------   ---------"
290 FOR I = 1 TO N
300 PRINT T(I),Q(I),S(I)
305 NEXT I
308 PRINT : PRINT : PRINT
310 PRINT "NEW DURATION = ":
320 INPUT D
324 PRINT "ORIGINAL DURATION =";
326 INPUT OD
330 FOR I = 1 TO 3
340 DS(I) = S(I)
350 NEXT I
360 FOR I = 4 TO N
370 DS(I) = S(I) - S(I-3)
380 NEXT I
385 FOR I = N+1 TO N+2
387 DS(I) = S(N) - S(I-3)
389 NEXT I
401 CLS
403 PRINT "NEW UNIT GRAPH"
404 PRINT : PRINT
405 PRINT " TIME          DISCHARGE"
406 PRINT " ------        ----------"
410 FOR I = 1 TO N+2
412 IF I >= N THEN T(I) = T(N) + (I-N) * DT
415 DS(I) = DS(I) * OD / D
417 DS(I) = INT ( DS(I) * 100 + .5 ) / 100
420 PRINT T(I),DS(I)
430 NEXT I
440 END


RUN



DEVELOPMENT OF S CURVE
```

```
NUMBER OF ORDINATES = ? 8


TIME INCREMENT = ? 4


TIME  1  =? 0
DISCHARGE  1  =? 0

TIME  2  =? 4
DISCHARGE  2  =? 20

TIME  3  =? 8
DISCHARGE  3  =? 50
```

```
TIME  4  =? 12
DISCHARGE  4  =? 70

TIME  5  =? 16
DISCHARGE  5  =? 65

TIME  6  =? 20
DISCHARGE  6  =? 60

TIME  7  =? 24
DISCHARGE  7  =? 40

TIME  8  =? 28
DISCHARGE  8  =? 0
```

TIME	DISCHARGE	S VALUE
0	0	0
4	20	20
8	50	70
12	70	140
16	65	205
20	60	265
24	40	305
28	0	305

NEW UNIT GRAPH

TIME	DISCHARGE
0	0
4	6.67
8	23.33
12	46.67
16	61.67
20	65
24	55
28	33.33
32	13.33
36	0

NEW DURATION = ? 12
ORIGINAL DURATION =? 4

Program notes

(1) Lines 0-305 — as per 'S' curve program.
(2) Lines 310-326 — enter new and original durations.
(3) Lines 330-350 — initial ordinal differences are equal to 'S' curve ordinates.
(4) Lines 360-380 — calculate main ordinal differences.
(5) Lines 385-389 — calculate final ordinal differences.
(6) Lines 401-405 — clear screen and print titles.
(7) Lines 410-430 — calculate ordinates of hydrograph, rounds-off values to two decimal places and print values.

Example 7.4 SYNHYD: development of surface runoff hydrograph from unit hydrograph and composite storm

The data provided below give details of a 4 h unit hydrograph and of a design storm from which all losses have been abstracted. Determine the hydrograph of surface runoff which will result from this composite storm.

Unit hydrograph data

Time (h)	0	2	4	6	8	10	12	14	16
Discharge (m³/s)	0	40	90	135	130	120	80	40	0

Storm data

Time (h)	0-4	4–8	8–12
Fall (cm)	0.4	1.2	0.5

The ordinates of the unit hydrograph must be multiplied by the depths of excess rainfall to provide three hydrographs of surface runoff, one for each of the intensities given in the design storm. These must then be superimposed with the correct time lag and added to give the composite surface runoff hydrograph.

```
10 CLS
20 PRINT "SYNTHETIC HYDROGRAPH"
30 PRINT "--------------------"
35 PRINT : PRINT
40 PRINT "FOR UNIT HYDROGRAPH"
50 INPUT "STATE NUMBER OF ORDINATES "; N
55 PRINT : PRINT
60 INPUT "STATE TIME INCREMENT "; DT
65 PRINT
70 INPUT "DURATION OF HYDROGRAPH = "; D
75 PRINT : PRINT
80 X = N + 2 * D / DT
85 DIM P(X),Z(X),SM(X),Q(X),T(X)
100 FOR I = 0 TO N - 1
105 U = I * DT
110 T(I) = 0 + U
120 PRINT "DISCHARGE AT TIME "; U ;" = ";
130 INPUT Q(I)
135 PRINT
140 NEXT I
150 CLS
160 PRINT "UNIT HYDROGRAPH DATA"
170 PRINT "--------------------"
175 PRINT
180 PRINT "TIME      DISCHARGE"
190 PRINT "-----    -----------"
195 PRINT
200 FOR I = 1 TO N - 1
210 PRINT T(I),Q(I)
220 NEXT I
225 PRINT : PRINT
230 INPUT " IS THIS CORRECT ( Y/N ) "; A$
240 IF A$ = "Y" THEN GOTO 300
250 CLS : GOTO 100
300 FOR I = 0 TO N + 2 * D / DT
310 P(I) = 0
320 Z(I) = 0
330 NEXT I
400 FOR J = D / DT TO N + D / DT
410 P(J) = Q( J - D / DT )
420 NEXT J
500 FOR K = 2 * D / DT TO N + 2 * D / DT
510 Z(K) = Q( K - 2 * D / DT )
520 NEXT K
610 PRINT : PRINT
620 INPUT "FIRST RAIN = "; R1
625 PRINT
630 INPUT "SECOND RAIN = "; R2
635 PRINT
640 INPUT "THIRD RAIN = "; R3
650 FOR I = 0 TO N + 2 * D / DT
655 T(I) = 0 + I * DT
```

```
660 SM(I) = R1 * Q(I) + R2 * P(I) + R3 * Z(I)
670 NEXT I
680 CLS
700 PRINT "COMPOSITE HYDROGRAPH DETAILS"
710 PRINT "---------------------------"
720 PRINT
730 PRINT "TIME        DISCHARGE"
740 PRINT "-----       -----------"
750 FOR I = 0 TO N + 2 * D / DT
760 PRINT T(I), SM(I)
770 NEXT I
780 END
```

RUN

SYNTHETIC HYDROGRAPH

FOR UNIT HYDROGRAPH
STATE NUMBER OF ORDINATES ? 9

STATE TIME INCREMENT ? 2

DURATION OF HYDROGRAPH = ? 4

UNIT HYDROGRAPH DATA

TIME	DISCHARGE
2	40
4	90
6	135
8	130
10	120
12	80
14	40
16	0

IS THIS CORRECT (Y/N) ? Y

FIRST RAIN = ? .4

SECOND RAIN = ? 1.2

THIRD RAIN = ? .5

DISCHARGE AT TIME 0 = ? 0

DISCHARGE AT TIME 2 = ? 40

DISCHARGE AT TIME 4 = ? 90

DISCHARGE AT TIME 6 = ? 135

DISCHARGE AT TIME 8 = ? 130

DISCHARGE AT TIME 10 = ? 120

DISCHARGE AT TIME 12 = ? 80

DISCHARGE AT TIME 14 = ? 40

DISCHARGE AT TIME 16 = ? 0

COMPOSITE HYDROGRAPH DETAILS

TIME	DISCHARGE
0	0
2	16
4	36
6	102
8	160
10	230
12	233
14	227.5
16	161
18	108
20	40
22	20
24	0
26	0

Program notes

(1) Lines 0-35 — clear screen and print title.
(2) Lines 40-75 — enter basic hydrograph data.
(3) Line 80 — determines array sizes.
(4) Line 85 — dimensions arrays.
(5) Lines 100-140 — enter hydrograph ordinates.
(6) Lines 150-240 — display entered data and allow modification if errors are present.
(7) Line 250 — clears screen and jumps back if error declared.
(8) Lines 300-330 — set all arrays to zero.
(9) Lines 400-520 — set up two arrays for the lagged data corresponding to storm periods 2 and 3.
(10) Lines 610-640 — enter details of storm.
(11) Lines 650-670 — add lagged ordinates and calculate time at each.
(12) Lines 680-780 — clear screen, print titles and calculated data, end program.

PROBLEMS

(7.1) The unit hydrograph shown below resulted from a 4 h duration storm. Write a program to determine the ordinates of the associated 'S' curve.

Time	0	1	2	3	4	5	6	7	8	9	10	11	12
Flow	0	11	31	50	70	68	63	52	42	33	27	23	18

Time	13	14	15	16	17	18	19	20	21	22	23	24
Flow	15	13	11	9	7	6	4	3	2	1	1.5	0

(7.2) Using the program developed in Problem (7.1), develop an additional routine to calculate the ordinates of a 3 h duration unit graph.

(7.3) Using the data of Example 7.2, write a program suitable for determining the ordinates of unit hydrographs for any integer multiple duration. Use the procedure of hydrograph lagging (see Section 7.3) and compare your results with those obtained using the program MODUR.

(7.4) The data below provide details of the runoff obtained from a 2 h duration storm which fell on a catchment area of 20^2 km.

Time (h)	0	2	4	6	8	10	12
Flow (m³/s)	60	245	175	132	100	75	60

Using this information, determine the ordinates of the unit graph and hence obtain the runoff which might be expected from a storm in which 2.5 cm is delivered in the first two-hour period and 3.0 cm is delivered in the second two-hour period (i.e. hours 2-4).

Chapter 8

Reservoir design

ESSENTIAL THEORY

8.1 Introduction

Reservoirs are designed essentially to take care of fluctuations in the demand or supply of water. Figure 8.1 shows a typical variation in demand for water in an urban community during a 24 h period. Relatively little water is used in the early hours of the morning with peak demands occurring around lunchtime and dinner in the evening. Similar variations occur if the demand for water is looked at on an annual basis. In general, relatively little water is used in the winter relative to that required during the summer when consumers use more water for baths, washing and watering lawns, etc. However, pipes carrying water to consumers must be designed to carry the maximum instantaneous demand rather than the average.

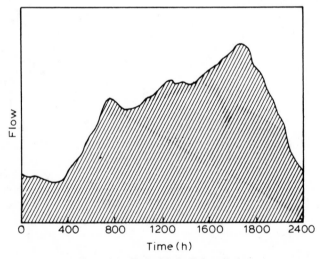

Figure 8.1 Typical daily flow variation

Depending on the degree of fluctuation about the norm, the maximum hourly demand, for example, may be as much as 1.5-2.0 times greater than the mean daily demand. Obviously, it would be very costly to run pipes of this size from the storage, or impounding, reservoir all the way to the consumer.

The solution to the problem of high cost lies in the use of two reservoirs. The main storage reservoir may be located at relatively large distances from the consumer. It is designed generally to provide long-term storage and to balance out the fluctuations in river flow over a period of a year or more. With suitable storage the pipe leading away from the impounding reservoir can be designed for a nearly constant average flow leading to a service reservoir placed fairly close to consumers. The service reservoir normally contains up to one day's storage and may be used to balance the fluctuation in the demand so that it operates with a relatively constant inflow and a fluctuating outflow. At the same time it provides a safeguard to allow for maintenance or to give an extra supply in the case of fire. The situation is shown diagrammatically in Figure 8.2.

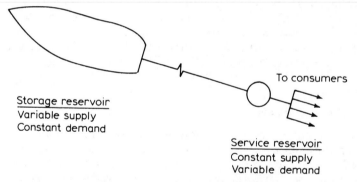

Figure 8.2 Storage and service reservoirs

Storage calculations for reservoir design involve the balancing of supply and demand so that stored water is available at times when the outflow exceeds the inflow and so that space is available for storage when the inflow is greater than the outflow. As indicated previously, a fairly simple design for an impounding reservoir would be based on a constant outflow and a varying inflow. Similar calculations for a service reservoir, however, would deal with a constant inflow and varying outflow.

Figure 8.3 shows various zones of storage in a reservoir. Dead storage, lying below the lowest outfall, and surcharge storage, lying

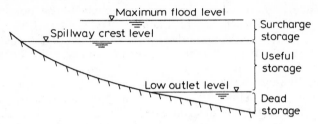

Figure 8.3   Primary storage zones

above spillway crest level, cannot be used and storage calculations refer essentially to the useful storage. Although in many cases it may be assumed that the surface level is horizontal, this will not necessarily be the case. For example, the surface level eighty miles upstream of the Wheeler Dam on the Tennessee River can vary by as much as 12 ft due largely to the increase in frictional resistance with increasing flow.

Storage calculations deal with the determination of the required reservoir capacity. However, in addition to storing water in times of shortfall, it is important that the dam be safe against flooding. This requires the design and construction of an adequate spillway. When the design flood has been chosen, or calculated, it is necessary to determine the effect that the reservoir will have on the shape and magnitude of the design flood as it moves through the reservoir to the spillway. This process is known as flood routing.

8.2   Storage calculations

A mass curve of supply is a curve showing the total (cumulative) volume entering a reservoir site over a certain time period, usually of years. Records are examined for critical dry periods and the mass curve may be drawn for two, three or more years. Flow data at monthly increments are usually sufficient but this depends on the scope of the project.

In the mass curve shown in Figure 8.4, the ordinate at any time, or the difference in ordinates between two times, provides a measure of the volume entering the reservoir. The full ordinate gives a measure of the total volume entering since time zero and the difference in ordinates a measure of the volume entering in the time period between the two ordinates. Since the plot is of volume against time, the slope of the curve at any time is equal to the rate at which water enters the reservoir, i.e. the rate of flow.

The difference between inflow and outflow may be compared by plotting the mass curve of demand on the same diagram as the mass

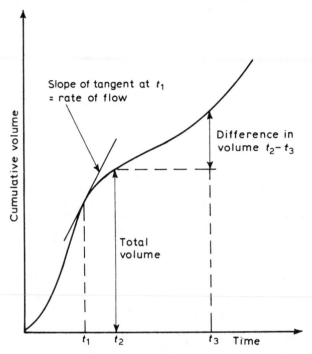

Figure 8.4   Simple mass curve

curve of inflow. For example, Figure 8.5 compares the situation in which the demand is constant and the inflow varies. It is assumed here that the reservoir is full at time zero. From point 0 to point A, the rate of inflow is greater than the demand rate and, because the reservoir is full at 0, the excess water must spill. The total volume of spill is given by the difference in the ordinates of the demand curve and supply curve. To compare the demand and supply rates after point A it is convenient to replot the demand curve starting at the point A. Then it will be seen that from A to D the demand exceeds the supply. The difference must be met from storage and the volume to be supplied from storage will be the ordinal difference at D. From D to E the supply again exceeds the demand. The reservoir will begin to fill and will be full at the point E. Thereafter, so long as the rate of supply exceeds the rate of demand, the reservoir will remain full and any excess water will spill. This is the situation which occurs between points E and B. At B it is again convenient to replot the demand curve so as to show the difference in the rates and in the ordinates. At point B the reservoir is full and, with a demand rate higher than the

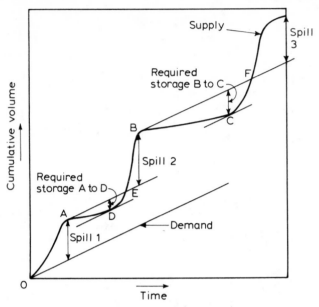

Figure 8.5   Details of mass curve

rate of supply, water must again be drawn from storage. Drawdown continues over the full time period from B to C. The ordinal difference at C again provides the volume to be drawn from storage. From C to F the supply exceeds the demand, the reservoir will fill up again, be full at F and will begin to spill.

Figure 8.5 shows three periods during which water spills; O to A, E to B, and F +. The addition of these ordinates gives the total volume spilled over the entire time period. Similarly, there are two occasions on which water is drawn from storage, i.e. from A to D and from B to C. The storage volume required during the first shortfall is given by the ordinal difference at D and that required during the second shortfall by the ordinal difference at C. The maximum storage required during the whole time period is therefore the maximum of these two ordinal differences. For the curves shown in Figure 8.5 this would be the ordinal difference at C. Note that the differences are not added because the reservoir fills up between the two drawdown periods.

The estimation of the safe yield for a given fixed reservoir capacity reverses the process. Here the storage capacity is plotted as an ordinate in the appropriate position and lines are drawn tangential to the mass curve to give the maximum possible demand rate. For the curves shown in Figure 8.6, the maximum demand which can be

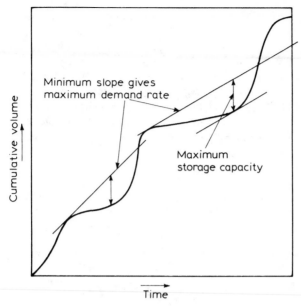

Figure 8.6 Safe yield for given capacity

achieved with the specified storage capacity is given by the slope of the tangent in the second of the two dry periods.

Care must be taken to specify the supply and demand accurately. Although these consist primarily of inflow and outflow, evaporation losses and downstream releases to satisfy riparian rights may constitute additional demands on the reservoir. Similarly, precipitation falling on the reservoir surface adds to the total supply of water.

The method can be illustrated most suitably by the use of a numerical example.

Data given below provide details of the flow entering a tropical river over a seventeen-month period (assume each month equals 30 days). The problem is to determine the reservoir capacity required to satisfy a constant demand of 72 m³/s over this time period. All figures have

Flow (cumec months) in 17 (30 day) months											
Apr.	May	June	July	Aug.	Sept.	Oct.	Nov.	Dec.	Jan.	Feb.	Mar.
8.8	18.7	79.3	104.9	240.9	249.4	221.0	65.2	28.3	22.7	1.7	8.5
41.7	71.1	147.6	218.8	164.1							

been adjusted to take account of evaporation, precipitation and losses, etc.

The mass curve is drawn in Figure 8.7 and volumes are specified in terms of cumec months. This is the volume which is obtained by a flow of 1 m³/s running for one month (i.e. $2.592 \times 10^6 \, \text{m}^3$). Plotting the cumulative supply and demand curves as indicated previously shows two dry periods from the beginning of April to mid June in the first year and from October to the end of May in the second year. The maximum deficit to be met from storage is 260 cumec months

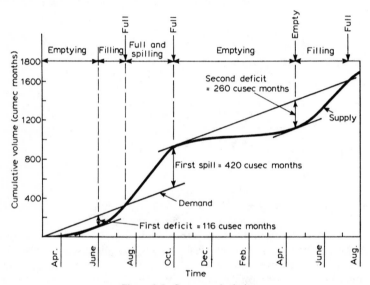

Figure 8.7   Storage calculation

and the total volume spilled equals 420 + cumec months. (There is a small spill at the end of the time period but the graph is not sufficiently accurate to determine the amount). The conditions of the reservoir at any time are shown along the top of Figure 8.7.

An alternative method of analysis involves keeping track of the excess and the deficit throughout the time period under consideration. Calculations are facilitated by a tabular approach such as that shown in Table 8.1. Here the data of the previous example have been re-used and the first three columns of Table 8.1 show the time of year, the supply and the demand. Again volumes are shown in cumec months. Comparisons between supply and demand on a monthly basis provide details of the deficits which must be met from storage or the excesses which either fill the reservoir or

Table 8.1

Month	Supply	Demand	Deficit	Excess	Volume
Apr.	8.8	72	63.2		−63.2
May	18.7	72	53.3		−116.5
June	79.3	72	Σ116.5	7.3	−109.2
July	104.9	72		32.9	−76.3
Aug.	240.9	72		168.9	full
Sept.	249.4	72		177.4	full
Oct.	221.0	72		149.0	full
Nov.	65.2	72	6.8	Σ535.5	−6.8
Dec.	28.3	72	43.7		−50.5
Jan.	22.7	72	49.3		−99.8
Feb.	1.7	72	70.3		−170.1
Mar.	8.5	72	63.5		−233.6
Apr.	41.7	72	30.3		−263.9
May	71.1	72	0.9		−264.8
June	147.6	72	Σ264.8	75.6	−189.2
July	218.8	72		146.8	−42.2
Aug.	164.1	72		92.1	full
				Σ314.5	

Notes

(1) Storage capacity required = 264.8 m³ month/s

(2) First spill (Aug.-Oct.) = 535.5−116.5 = 419.0 m³ month/s

Second spill (July-Aug. yr 2) = 314.5−264.8 = 49.7 m³ month/s

are spilled. These are shown in columns 4 and 5 of Table 8.1. Column 6 keeps a running record of the volume in the reservoir and is based on the assumption that the reservoir is full at the start of the time period. This running total is particularly valuable in situations where the reservoir does not completely refill between each dry period. Table 8.2, for example considers this type of situation and compares the supply and demand in a river over a three-year period.

Here it will be seen that there are three periods of excess flow and three periods in which the flow is less than the demand. However, after the first period of excess, there is only one brief period, from January to March of year three, when the reservoir is again full. This example assumes that the reservoir is full at the start of the time period considered. From October to April the excess water is spilled and from April of year one to August of year two the reservoir level is drawn down. The maximum deficit occurs in August and is equal to 49 acre ft. In September of year two, a period of excess begins and the reservoir begins to fill up becoming full around the beginning of January in year three. The remainder of the excess in January and all the excess in February and March again spills giving a total spill of 103 + 61 = 164 acre ft over the three-year time period. If the reservoir were completely empty at the beginning of the time period,

Table 8.2

Time	Flow (acre ft per mth)	Demand (acre ft per mth)	Excess (acre ft)	Deficit (acre ft)	Volume (acre ft)	Spill (acre ft)
Oct.	18	5	13		full	13
Nov.	22	5	17		full	17
Dec.	17	5	12		full	12
Jan.	26	5	21		full	21
Feb.	15	5	10		full	10
Mar.	32	5	27		full	27
Apr.	8	5	3		full	3
May	3	5		2	−2	Σ103
June	0	5		5	−7	
July	0	5		5	−12	
Aug.	0	5		5	−17	
Sept.	0	5		5	−22	
Year 2						
Oct.	5	5	0	0	−22	
Nov.	6	5	1		−21	
Dec.	6	5	1		−20	
Jan.	5	5	0	0	−20	
Feb.	3	5		2	−22	
Mar.	2	5		3	−25	
Apr.	1	5		4	−29	
May	0	5		5	−34	
June	0	5		5	−39	
July	0	5		5	−44	
Aug.	0	5		5	−49	
Sept.	7	5	2		−47	
Year 3						
Oct.	15	5	10		−37	
Nov.	17	5	12		−25	
Dec.	25	5	20		−5	
Jan.	47	5	42		full	37
Feb.	16	5	11		full	11
Mar.	18	5	13		full	13
Apr.	4	5		1	−1	Σ61
May	4	5		1	−2	
June	0	5		5	−7	
July	1	5		4	−11	
Aug.	3	5		2	−13	
Sept.	4	5		1	−14	

Notes
(1) Storage capacity required = 49 acre ft
(2) Total spill = 164 acre ft

then 49 acre ft. of the initial 103 excess would be used to fill the reservoir. The reservoir would fill around the beginning of January and the remainder of that initial excess (i.e. $103 - 49 = 54$ acre ft.) would be spilled. Thereafter the sequence of events would be identical to that described earlier. However, in this case, the total spill would be $54 + 61 = 115$ acre ft.

The tabular method of analysis is of course ideally suited to computational procedures.

8.3 Flood routing

Flood routing is a technique used to compute the effect of system storage and system dynamics on the shape and movement of a flood wave. When, for example, the unit hydrograph has been developed and used to predict the hydrograph of flow at the exit of a drainage basin a flood routing procedure could be used to calculate the shape of the flood hydrograph at a downstream point and the time at which the flood would occur. The effect of the system through which the flood passes is to cause the peak flow to be reduced and the time of occurrence of the peak flow to be delayed. These effects are termed attenuation and lag respectively and are shown in Figure 8.8.

Two methods of flood routing are generally used. The hydraulic

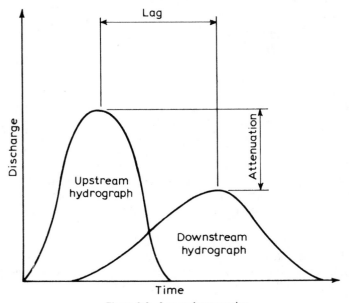

Figure 8.8 Lag and attenuation

method involves the integration of the partial differential equations for unsteady varied flow whereas hydrologic methods, which will be considered here, are essentially empirical, arithmetic, procedures relating inflow, outflow and storage.

All hydrologic methods are based on the fundamental storage equation

$$I-O = \frac{dS}{dt} \tag{8.1}$$

where I = rate of inflow, O = rate of outflow, S = storage volume and t = time. Because I, O and S vary with time, it is possible to convert this equation to one involving finite differences, i.e.

$$\frac{(I_1 + I_2)}{2} \Delta t - \frac{(O_1 + O_2)}{2} \Delta t = S_2 - S_1 \tag{8.2}$$

where Δt = time increment between times indicated by subscripts 1 and 2.

The purpose of a reservoir routing calculation is generally to assess the adequacy of the design spillway to pass the outflow caused by the entry of the design flood to the reservoir. The general procedure is to assume a spillway design (size, shape, elevation, etc.) and to check that design to ensure that the flood can pass without unacceptable high water elevations. The calculation is based on Equation (8.2) with some rearrangement. All known variables are taken to the left-hand side of the equation and unknown variables are brought to the right-hand side. Thus

$$\frac{(I_1 + I_2)}{2} \Delta t - \frac{O_1 \Delta t}{2} + S_1 = \frac{O_2 \Delta t}{2} + S_2 \tag{8.3}$$

It is assumed that full knowledge is available for the inflow hydrograph and that, at the start of the calculation, the elevation of the water level H and, therefore, the initial outflow over the spillway crest, is known. Similarly, with a known initial elevation, the initial storage volume in the reservoir is known. The solution is advanced in time increments of Δt and the basic problem is to use Equation (8.3) to calculate values of O_2 and S_2. These represent outflow and storage at the end of the first time increment and can therefore be used as initial values for calculations involving the second time increment from t_2 to t_3.

A technique which is partially graphical and partially arithmetic is normally employed. For any particular simple spillway, the discharge O depends on the head over the spillway. For example, with a weir spillway, the discharge is given by

$$O = CLH^{3/2} \tag{8.4}$$

where C = constant, L = length of crest and H = head over crest. Similarly, for any reservoir shape, the storage volume in the reservoir may be related to elevation. In the simplest case of a reservoir having vertical sides, the relationship between elevation and storage will be a straight line.

The relationships discussed above are shown in Figure 8.9 (a–c). For the solution of Equation (8.3) it is convenient to construct another diagram using the elevation discharge relationship (b) and the elevation storage relationship (c). Once a time increment has been chosen for the calculations, values of O and S may be obtained for any elevation H. It is thus possible to determine a relationship between elevation and $O\Delta t/2 + S$. This relationship is shown diagrammatically in Figure 8.9 (d).

The analysis is now fairly straightforward. Having chosen a value for Δt, suitable for calculation, and knowing the initial elevation of the reservoir surface, the values of all parameters on the left-hand side of Equation (8.3) may be obtained from curves similar to those drawn in Figure 8.9 (a–c). Here it is worth mentioning that because the analysis deals with change in storage rather than absolute storage (see Equation (8.2)), it is possible to assign zero storage arbitrarily to

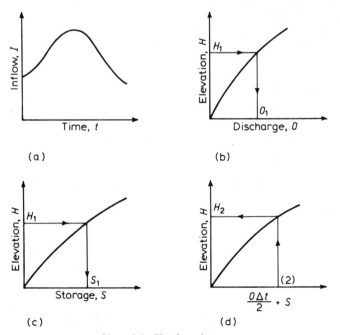

Figure 8.9 Flood routing curves

any elevation below the spillway crest. The calculation is repetitive and the procedure is as follows.

(1)   Choose Δt and draw Figure 8.9(d) using Figure 8.9 (b–c).
(2)   At time t_1, knowing H_1, determining I_1, S_1, O_1 and I_2 at time $(t_1 + \Delta t)$, from Figure 8.9 (a–c).
(3)   Calculate $[O_2 \Delta t/2 + S_2]$ from Equation (8.3) using values determined in (2).
(4)   From Figure 8.9(d) determine H_2. This is the elevation at the end of the first time increment and at the beginning of the second.
(5)   Repeat the calculation starting at time t_2 with known value of H_2. Continue to later times as necessary.

The following example can be used to illustrate the method more completely.

A reservoir having a constant surface area of 100 acres has been constructed with two separate uncontrolled spillways having crest levels of datum $+98.0$ ft and datum $+96.0$ ft. The respective widths are 100 ft and 30 ft. Initially the water surface is at an elevation of datum $+98.4$ ft. Determine the maximum discharge over each spillway resulting from the inflow hydrograph given below. The outflow in ft^3/s is given by $0 = 3.3 \, LH^{3/2}$.

Time (h)	0	1	2	3	4	5
Inflow (Acre ft/h)	40	60	100	110	100	40

The total outflow from both spillways is given by

$$O = 330 \, (H{-}98)^{3/2} + 99 \, (H{-}96)^{3/2} \text{ ft}^3/\text{s} \tag{8.5}$$

where the parts in brackets are zero if negative.
Assigning zero storage to elevation datum $+96.0$ ft,

$$S = 100(H{-}96.0) \tag{8.6}$$

These equations are now sufficient for the development of the three

Table 8.3

H	O	S	S + OΔT/2
			Δt = 1 hr
(ft)	(Acre ft/h)	(Acre ft)	(Acre ft)
96	0	0	0
97	8.2	100	104.1
98	23.4	200	211.7
99	69.8	300	334.9
100	142.5	400	471.3

graphs necessary for the calculation. Tabulated data are shown in Table 8.3. This has been prepared on the assumption of a one hour time increment. These data, together with the inflow hydrograph, are plotted in Figures 8.10 to 8.13. (Here it may be noted that not all of these curves are necessary in all cases. In this fairly simple

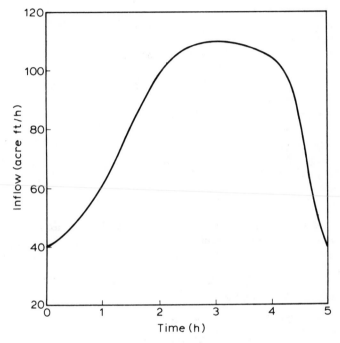

Figure 8.10   Inflow hydrograph

example, inflow data are given at hourly figures and may be abstracted from the data table given whereas outflow and storage data may be obtained directly from Equations (8.5) and (8.6). In more complicated cases however, these may be replaced by empirical relationships and the use of a graphical approach will then be necessary.)

The full calculation is shown in Table 8.4. The calculation starts at time 0 with an inflow of 40 acre ft/h and an initial elevation of 98.4 ft. The values of the variables given in columns 3 to 6 are obtained from Figures 8.10 to 8.12 (or the given hydrograph and Equations (8.5) and (8.6)). Column 7 is obtained by the substitution of these values into Equation (8.3) and the value of H_2 in column 8 is

Table 8.4

Time (h)	I $\left(\dfrac{\text{Acre ft}}{\text{h}}\right)$	$\left(\dfrac{I_1+I_2}{2}\right)\Delta t$ (Acre ft)	H_1 (ft)	S_1 (Acre ft)	O_1 $\left(\dfrac{\text{Acre ft}}{\text{h}}\right)$	$S_2+\dfrac{O_2\Delta T}{2}$ (Acre ft)	H_2 (ft)
0	40						
(Increment No. 1)		50	98.4	240	37.3	271.4	98.5
1	60						
(Increment No. 2)		80	98.5	250	42.0	309.0	98.8
2	100						
(Increment No. 3)		105	98.8	280	57.8	356.1	99.15
3	110						
(Increment No. 4)		105	99.15	315	79.4	380.3	99.35
4	100						
(Increment No. 5)		70	99.35	335	92.95	358.4	99.3
5	40						

Note 1 Acre ft/h = 12.1 ft^3/s

obtained from Figure 8.13. That elevation is the starting elevation for the second time increment, running from hour one to hour two, and the process is repeated as necessary.

From Table 8.4 it is seen that the maximum water level of 99.35 ft above datum occurs at hour four. The discharge from the two spillways is then given by

$$O_1 = 330(99.35-98)^{1.5} = 517.6 \text{ ft}^3/\text{s} \tag{8.7}$$

$$O_2 = 99(99.35-96)^{1.5} = 607.0 \text{ ft}^3/\text{s} \tag{8.8}$$

This gives a total discharge of 1124.6 ft^3/s.

Reservoir routing is fairly simple because it is assumed that the discharge and storage are directly related to the downstream surface elevation. This is not necessarily the case, however, and is certainly not true in rivers where prism storage and wedge storage, as indicated in Figure 8.14, are both present. In that situation, storage may be expressed in terms of the inflow and the discharge. Then, under these conditions

$$S = K(xI + (1-x)0) \tag{8.9}$$

where x is a dimensionless constant and K is a storage constant. Methods of determining values of K and x are described elsewhere. However, approximate values are as follows. The storage constant K is roughly equal to the time taken by the flood wave in travelling over

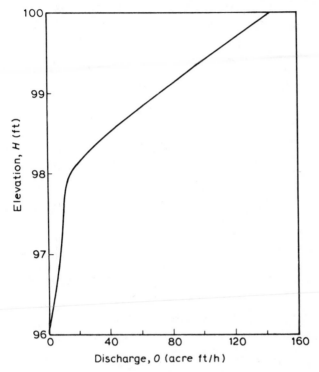

Figure 8.11   Spillway discharge characteristic

the reach of the river chosen for routing. Values of x can range from 0 to 0.5 but usually lie between 0.2 and 0.3. There is an added requirement, namely that the time step must be chosen so that

$$2Kx < \Delta t < K \tag{8.10}$$

The procedure for routing a flood from one point to another follows that described earlier but uses Equation (8.9) instead of the basic storage equation (i.e. one similar to (8.6)). Similarly, it is necessary to modify the discharge equation (i.e. Equation (8.5)) and this can be done by combining Equations (8.9) and (8.2).

One method of determining K and x is to estimate initial values and to use these to route a known flood between two points at which field measurements were made. Success can be gauged by comparing the results of the routing calculation with the measured floods at the two points. The comparison of the actual with the measured provides guidance to modify the chosen values of K and x. If the routed flood is attenuated more than the measured flood, the value

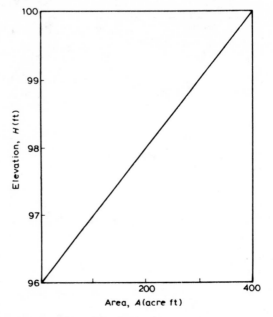

Figure 8.12   Reservoir storage curve

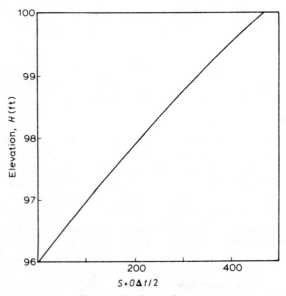

Figure 8.13   Composite curve

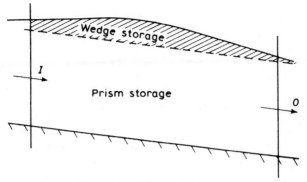

Figure 8.14 Prism and wedge storage

of x should be increased. If there is too much lag, the value of K should be decreased.

WORKED EXAMPLES

Example 8.1 CAPACITY: estimation of required storage volume

The data given below refer to conditions at a proposed reservoir site. Assume that the surface area will increase by 1000 acres and that 25% of the rain falling on the land area to be flooded reached the stream prior to reservoir construction. Evaporation has been measured using a pan with a coefficient equal to 0.7. Determine the required useful storage.

The solution must compare demand and supply in order to evaluate deficit, excess and a total running volume as indicated in

Month (1)	Flow (2) (Acre ft)	Precipitation (3) (In)	Demand (4) (Acre ft)	Evaporation (5) (In)	Compensation flow (6) (Acre ft)
Jan.	2000	4.6	45	3.4	80
Feb.	4500	4.8	38	5.1	80
Mar.	35	0.8	80	5.7	35
Apr.	12	0.7	125	6.2	12
May	4	0.2	150	5.5	4
June	3	0.1	150	4.5	3
July	1	0	120	2.9	1
Aug.	0	0	170	1.8	0
Sept	0	0	90	0.7	0
Oct.	0	0	35	0.9	0
Nov.	0	0.7	25	0.9	0
Dec.	4	3.5	25	2.6	4

Tables 8.1 and 8.2. The basic supply and demand in this problem are given in columns 2 and 4. However, the compensation flow, shown in column 6, represents an additional demand. This is water which must be released downstream to meet prior water rights. In this problem it has been assumed that the compensation flow will be either 80 acre ft per month or the flow in the river whichever is least. If 25% of the rain falling on the land area to be flooded (i.e. 1000 acres) reached the stream in the past, then 25% of the precipitation appears as part of the flow in column 2. The increase in the water entering the reservoir after construction with the rain falling on the whole reservoir surface will then by 75% of that shown in column 3; i.e. the volume given by

$$\text{precipitation volume} = p \times 1000 \times 0.75/12 \text{ Acre ft} \qquad (8.11)$$

If it is assumed that the lake evaporation will be $0.7 \times$ pan evaporation, the abstraction of water (additional demand) caused by evaporation is given by

$$\text{Evaporation volume} = E \times 1000 \times 0.7/12 \text{ Acre ft} \qquad (8.12)$$

Using these equations and the data given, it is possible to determine the total demand in any month. This is given by the summation of the demand, the compensation flow and evaporation volume. The total supply may be similarly evaluated as the summation of flow and precipitation volume. These may then be compared in the normal manner.

```
10 REM  ESTIMATION OF REQUIRED STORAGE VOLUME - RESERVOIR FULL AT TIME 0
20 CLS
30 PRINT "CAPACITY - ESTIMATION OF REQUIRED STORAGE VOLUME"
40 PRINT "---------------------------------------------------"
50 PRINT : PRINT : PRINT
60 PRINT "ENTER NUMBER OF PERIODS ";
70 INPUT N
80 DIM P$(N),FS(N),PI(N),PA(N),FD(N),EI(N)
90 DIM EA(N),CW(N),VE(N),VD(N),VT(N+1),SP(N)
100 CLS
110 PRINT "INPUT DATA FOR CAPACITY "
120 PRINT "------------------------"
130 PRINT : PRINT : PRINT
140 PRINT TAB(1);"PERIOD";TAB(21);"SUPPLY";TAB(49);"DEMAND"
150 PRINT TAB(1);"------";TAB(21);"------";TAB(49);"------"
160 PRINT TAB(16);"FLOW";TAB(29);"PRECIP.";TAB(44);"FLOW";TAB(56);"EVAP."
170 PRINT "-------------------------------------------------------------"
180 PRINT
190 FOR I = 1 TO N
200 READ P$(I),FS(I),PI(I),FD(I),EI(I)
210 PRINT P$(I),FS(I),PI(I),FD(I),EI(I)
220 NEXT I
230 PRINT
240 PRINT "PRESS RETURN KEY  < enter > TO CONTINUE ";
250 INPUT A$
260 FOR I = 1 TO N
270 PA(I) = CINT ( PI(I) * 1000 * ( .75 / 12 ) )
280 EA(I) = CINT ( EI(I) * 1000 * ( .7  / 12 ) )
290 A1 = FS(I)
```

```
300 IF A1 > 80 THEN CW(I) = 80 : GOTO 310
305 CW(I) = A1
310 NEXT I
320 VT(0) = 0
330 FOR I = 1 TO N
340 S = FS(I) + PA(I)
350 D = FD(I) + EA(I) + CW(I)
360 T = S - D
370 IF T > 0 THEN VE(I) = T
380 IF T <= 0 THEN VD(I) = T
390 VT(I) = VT( I-1 ) + T
400 IF VT(I) > 0 THEN SP(I) = VE(I)
410 IF VT(I) > 0 THEN VT(I) = 0
420 NEXT I
430 REM  OUTPUT RESULTS
440 CLS
450 LPRINT "ESTIMATION OF REQUIRED STORAGE VOLUME"
460 LPRINT "-------------------------------------"
470 LPRINT
480 LPRINT "NOTE : Reservoir full at time 0 "
490 LPRINT : LPRINT
500 LPRINT TAB(1);"PERIOD";TAB(22);"SUPPLY";TAB(45);"DEMAND";TAB(85);"VOLUME"
510 LPRINT TAB(1);"------";TAB(22);"------";TAB(45);"------";TAB(85);"------"
520 LPRINT TAB(8);"FLOW" TAB(18);"PRECIPITATION" TAB(35);"FLOW" TAB(45);"EVAPORATION" TAB(62);"COMP. FLOW" TAB(76);
525 LPRINT "EXCESS" TAB(86);"DEFICIT" TAB(96);"VOLUME" TAB(106);"SPILL"
530 AF$ = "acre-ft" : IN$ = "inches"
540 LPRINT TAB(8);AF$;TAB(18);IN$;TAB(26);AF$;TAB(35);AF$;TAB(45);IN$;TAB(53);AF$;TAB(62);AF$;TAB(76);AF$;TAB(86);AF$;
545 LPRINT TAB(96);AF$;TAB(106);AF$
550 LPRINT "---------------------------------------------------------------------------------------------------------";
555 LPRINT "--------------------"
560 FOR I = 1 TO N
570 LPRINT TAB(1);P$(I);TAB(8);FS(I);TAB(18);PI(I);TAB(26);PA(I);TAB(35);FD(I);TAB(45);EI(I);TAB(53);EA(I);TAB(62);
575 LPRINT CW(I);TAB(76);VE(I);TAB(86);VD(I);TAB(96);VT(I);TAB(106);SP(I)
580 NEXT I
590 LPRINT : LPRINT : LPRINT
600 PRINT "IF YOU WISH TO DO THIS PROGRAM AGAIN, TYPE IN NEW DATA STATEMENTS AND RE-RUN."
610 END
620 :
630 DATA JAN,2000,4.6,45,3.4
640 DATA FEB,4500,4.8,38,5.1
650 DATA MAR,35,0.8,80,5.7
660 DATA APR,12,0.7,125,6.2
670 DATA MAY,4,0.2,150,5.5
680 DATA JUN,3,0.1,150,4.5
690 DATA JUL,1,0,120,2.9
700 DATA AUG,0,0,170,1.8
710 DATA SEP,0,0,90,0.7
720 DATA OCT,0,0,35,0.9
730 DATA NOV,0,0.7,25,0.9
740 DATA DEC,4,3.5,25,2.6


RUN

CAPACITY - ESTIMATION OF REQUIRED STORAGE VOLUME
------------------------------------------------

ENTER NUMBER OF PERIODS ? 12


INPUT DATA FOR CAPACITY
-----------------------
```

PERIOD	SUPPLY		DEMAND	
	FLOW	PRECIP.	FLOW	EVAP.
JAN	2000	4.6	45	3.4
FEB	4500	4.8	38	5.1
MAR	35	.8	80	5.7

APR	12	.7	125	6.2
MAY	4	.2	150	5.5
JUN	3	.1	150	4.5
JUL	1	0	120	2.9
AUG	0	0	170	1.8
SEP	0	0	90	.7
OCT	0	0	35	.9
NOV	0	.7	25	.9
DEC	4	3.5	25	2.6

PRESS RETURN KEY (enter) TO CONTINUE ?

ESTIMATION OF REQUIRED STORAGE VOLUME

NOTE : Reservoir full at time 0

PERIOD	SUPPLY			DEMAND				VOLUME			
	FLOW	PRECIPITATION	FLOW	EVAPORATION		COMP. FLOW	EXCESS	DEFICIT	VOLUME	SPILL
	acre-ft	inches acre-ft	acre-ft	inches acre-ft		comp.flow acre-ft	acre-ft	acre-ft	acre-ft	acre-ft
JAN	2000	4.6	288	45	3.4	198	80	1965	0	0	1965
FEB	4500	4.8	300	38	5.1	298	80	4384	0	0	4384
MAR	35	.8	50	80	5.7	333	35	0	-363	-363	0
APR	12	.7	44	125	6.2	362	12	0	-443	-806	0
MAY	4	.2	13	150	5.5	321	4	0	-458	-1264	0
JUN	3	.1	6	150	4.5	263	3	0	-407	-1671	0
JUL	1	0	0	120	2.9	169	1	0	-289	-1960	0
AUG	0	0	0	170	1.8	105	0	0	-275	-2235	0
SEP	0	0	0	90	.7	41	0	0	-131	-2366	0
OCT	0	0	0	35	.9	53	0	0	-88	-2454	0
NOV	0	.7	44	25	.9	53	0	0	-34	-2488	0
DEC	4	3.5	219	25	2.6	152	4	42	0	-2446	0

IF YOU WISH TO DO THIS PROGRAM AGAIN, TYPE IN NEW DATA STATEMENTS AND RE-RUN.

Program notes

(1) Lines 0-50 — clear screen and print title.
(2) Lines 60-90 — enter number of periods and dimension arrays.
(3) Lines 100-250 — read and print the input in the data statements.
(4) Lines 260-310 — calculate volumes for precipitation, evaporation and compensation flow.
(5) Lines 320-420 — get supply, demand and total: calculate running total, excess or deficit and spill volumes.
(6) Lines 430-590 — output results.
(7) Lines 600-620 — repeat program instructions.
(8) Lines 630-740 — data statements.

Example 8.2 ROUTING: flood routing through reservoirs

A steep-sided lake with a surface area of 500 acres discharges into a steep rectangular channel 25 ft wide. The system is initially steady-state with 1000 ft³/s entering the lake through a river at the upstream end and the same flow rate leaving via the spillway. Data given below provide details of a flood which then enters the river. Determine the outflow hydrograph over the 48 h period.

Time (h)	0	3	6	9	12	15	18	21
Flow (ft³/s)	1000	1200	1500	2000	2600	2900	3100	3000

Time (h)	24	27	30	33	36	39	42	45	48
Flow (ft³/s)	2800	2600	2300	2100	1700	1400	1100	1000	1000

The lake discharges to a steep channel and it may, therefore, be assumed that critical conditions exist at the outlet. Thus, assuming stillwater upstream, the discharge and elevation above the outlet crest are related by

$$H = 3/2 \left(\frac{O^2}{L^2 g} \right)^{1/3} \tag{8.13}$$

For a steady flow of 1000 ft³/s, Equation (8.13) gives the head as 5.52 ft.

Equation (8.13) is the basic outflow elevation equation and may be rewritten as

$$O = 77.22 \, H^{3/2} \, \text{ft}^3/\text{s} \tag{8.14}$$

If the steep sides are assumed to be vertical, then the storage equation is a straight line. Zero storage may be assigned to any level less than 5.52 ft above the crest and for convenience will be assigned to crest level (i.e. $H = 0$). Thus the storage is given by

$$S = 500 \, H \tag{8.15}$$

Because of the semi-graphical procedure which must be used, particularly in the estimation of elevation at the end of each time increment, a program will be developed to assist in the numerical calculations but this will still require the preparation of a graph similar to that in Figure 8.9 (d). The combined storage outfall graph will be determined from Equations (8.14) and (8.15). That graph will then be plotted manually and the program will continue to develop the tabular solution using the manually plotted curve together with Equation (8.3).

```
100 CLS
120 PRINT "   FLOOD ROUTING CALCULATION"
140 PRINT : PRINT
160 PRINT "ENTER HYDROGRAPH DATA"
```

```
180 PRINT
200 INPUT "NUMBER OF ORDINATES ="; N
220 DIM Q(N),T(N),QT(N),HD(N)
240 INPUT "ZERO TIME = "; T1
260 PRINT
280 INPUT "TIME INCREMENT = "; DT
300 FOR I = 1 TO N
320 T(I) = T1 + ( I - 1 ) * DT
340 PRINT "DISCHARGE AT TIME "; T(I) ;" = ";
360 INPUT Q(I)
380 NEXT I
390 CLS
400 PRINT "   INPUT HYDROGRAPH "
410 PRINT " ------------------"
440 PRINT "TIME        DISCHARGE"
460 FOR I = 1 TO N
480 PRINT T(I),Q(I)
500 NEXT I
520 PRINT :.PRINT
540 INPUT "IS THE INPUT CORRECT (Y/N) ?"; A$
560 IF A$ = "N" THEN GOTO 240
580 PRINT : PRINT
590 GOSUB 5000
600 FOR I = 1 TO N-1
620 PRINT " TIME INCREMENT NO "; I
640 PRINT "-------------------------"
660 PRINT "HEAD AT TIME "; I ;" =";
680 INPUT H
700 PRINT
720 GOSUB 6000
740 PRINT
760 PRINT
780 GOSUB 7000
790 PRINT
800 DS = (Q(I) + Q(I + 1)) * DT / 2 - D * DT / 2 + S
820 PRINT
840 PRINT " S+Q*DT/2 ="; DS
850 QT(I) = INT ( D * 100 + .5 ) / 100
855 HD(I) = H
860 PRINT
880 PRINT " FROM GRAPH DETERMINE HEAD AT"
900 PRINT " END OF INCREMENT"
920 PRINT " WHEN READY PRESS RETURN"
940 PRINT " AND ENTER HEAD WHEN REQUESTED"
960 PRINT
980 INPUT A$
1010 CLS
1015 IF I = N - 1 THEN GOSUB 8000
1020 NEXT I
1040 CLS
1060 PRINT "FINAL RESULTS"
1080 PRINT "-------------"
1120 PRINT "TIME    INFLOW    OUTFLOW  HEAD"
1160 FOR I = 1 TO N
1180 PRINT T(I);TAB(11);Q(I);TAB(18);QT(I);TAB(28);HD(I)
1190 NEXT I
2000 END
5000 CLS
```

```
5020 PRINT " SET UP ROUTING GRAPHS"
5040 PRINT "----------------------"
5060 PRINT
5080 PRINT "S+0*DT/2 AS FUNCTION OF HEAD"
5100 PRINT "---------------------------"
5120 PRINT
5140 PRINT "ENTER HEAD AS PROMPTED"
5160 PRINT
5180 PRINT "ENTER 10000 WHEN COMPLETE"
5200 PRINT
5220 INPUT " HEAD = "; H1
5230 IF H1 = 10000 THEN GOTO 5340
5240 ST = 500 * H1 + DT * 77.22 * H1 ^ 1.5 / 2
5280 PRINT "PLOT THE POINT"
5300 PRINT "FOR HEAD = "; H1 ;" S+0*DT/2 = "; ST
5310 PRINT
5320 GOTO 5220
5340 PRINT " IS THE GRAPH COMPLETE ( Y/N ) ";
5360 INPUT T$
5380 IF T$ = "N" THEN GOTO 5140
5400 CLS
5420 RETURN
6000 D = 77.22 * H ^ 1.5
6020 RETURN
7000 S = 500 * H
7020 RETURN
8000 PRINT
8020 PRINT " HEAD IN LAST INCREMENT = ";
8040 INPUT H
8060 GOSUB 6000
8080 HD(N) = H
8100 QT(N) = INT (D * 100 + .5) / 100
8120 RETURN


RUN



    FLOOD ROUTING CALCULATION


ENTER HYDROGRAPH DATA

NUMBER OF ORDINATES =? 17
ZERO TIME = ? 0

TIME INCREMENT = ? 3

DISCHARGE AT TIME  0  = ? 1000
DISCHARGE AT TIME  3  = ? 1200          DISCHARGE AT TIME  27  = ? 2600
DISCHARGE AT TIME  6  = ? 1500          DISCHARGE AT TIME  30  = ? 2300
DISCHARGE AT TIME  9  = ? 2000          DISCHARGE AT TIME  33  = ? 2100
DISCHARGE AT TIME  12  = ? 2600         DISCHARGE AT TIME  36  = ? 1700
DISCHARGE AT TIME  15  = ? 2900         DISCHARGE AT TIME  39  = ? 1400
DISCHARGE AT TIME  18  = ? 3100         DISCHARGE AT TIME  42  = ? 1100
DISCHARGE AT TIME  21  = ? 3000         DISCHARGE AT TIME  45  = ? 1000
DISCHARGE AT TIME  24  = ? 2800         DISCHARGE AT TIME  48  = ? 1000
```

```
 INPUT HYDROGRAPH
 ------------------
TIME       DISCHARGE
 0          1000
 3          1200
 6          1500
 9          2000
 12         2600
 15         2900
 18         3100
 21         3000
 24         2800
 27         2600
 30         2300
 33         2100
 36         1700
 39         1400
 42         1100
 45         1000
 48         1000
```

IS THE INPUT CORRECT (Y/N) ?? Y


```
 SET UP ROUTING GRAPHS
 -----------------------

 S+O*DT/2 AS FUNCTION OF HEAD
 ----------------------------

 ENTER HEAD AS PROMPTED

 ENTER 10000 WHEN COMPLETE

  HEAD = ? 5.0
 PLOT THE POINT
 FOR HEAD = 5  S+O*DT/2 = 3795.019

  HEAD = ? 6.0
 PLOT THE POINT
 FOR HEAD = 6  S+O*DT/2 = 4702.346

  HEAD = ? 7.0
 PLOT THE POINT
 FOR HEAD = 7  S+O*DT/2 = 5645.202

  HEAD = ? 8.0
 PLOT THE POINT
 FOR HEAD = 8  S+O*DT/2 = 6620.934

  HEAD = ? 9.0
 PLOT THE POINT
 FOR HEAD = 9  S+O*DT/2 = 7627.408

  HEAD = ? 10.0
 PLOT THE POINT
 FOR HEAD = 10  S+O*DT/2 = 8662.866

  HEAD = ? 11.0
 PLOT THE POINT
 FOR HEAD = 11  S+O*DT/2 = 9725.811

  HEAD = ? 12.0
 PLOT THE POINT
 FOR HEAD = 12  S+O*DT/2 = 10814.96

  HEAD = ? 10000
 IS THE GRAPH COMPLETE ( Y/N ) ? Y
```

```
TIME INCREMENT NO  1
--------------------------
HEAD AT TIME  1  =? 5.52




S+O*DT/2 = 4557.794

FROM GRAPH DETERMINE HEAD AT
END OF INCREMENT
WHEN READY PRESS RETURN
AND ENTER HEAD WHEN REQUESTED

?
```

```
TIME INCREMENT NO  4
--------------------------
HEAD AT TIME  4  =? 7.92




S+O*DT/2 = 8278.283

FROM GRAPH DETERMINE HEAD AT
END OF INCREMENT
WHEN READY PRESS RETURN
AND ENTER HEAD WHEN REQUESTED

?
```

```
TIME INCREMENT NO  2
--------------------------
HEAD AT TIME  2  =? 5.8




S+O*DT/2 = 5332.058

FROM GRAPH DETERMINE HEAD AT
END OF INCREMENT
WHEN READY PRESS RETURN
AND ENTER HEAD WHEN REQUESTED

?
```

```
TIME INCREMENT NO  5
--------------------------
HEAD AT TIME  5  =? 9.60




S+O*DT/2 = 9604.693

FROM GRAPH DETERMINE HEAD AT
END OF INCREMENT
WHEN READY PRESS RETURN
AND ENTER HEAD WHEN REQUESTED

?
```

```
TIME INCREMENT NO  3
--------------------------
HEAD AT TIME  3  =? 6.07




S+O*DT/2 = 6589.694

FROM GRAPH DETERMINE HEAD AT
END OF INCREMENT
WHEN READY PRESS RETURN
AND ENTER HEAD WHEN REQUESTED

?
```

```
TIME INCREMENT NO  6
--------------------------
HEAD AT TIME  6  =? 10.9




S+O*DT/2 = 10281.68

FROM GRAPH DETERMINE HEAD AT
END OF INCREMENT
WHEN READY PRESS RETURN
AND ENTER HEAD WHEN REQUESTED

?
```

```
TIME INCREMENT NO  7
--------------------------
HEAD AT TIME  7  =? 11.59
```

```
S+O#DT/2 = 10374.69

FROM GRAPH DETERMINE HEAD AT
END OF INCREMENT
WHEN READY PRESS RETURN
AND ENTER HEAD WHEN REQUESTED

?
```

```
TIME INCREMENT NO  10
--------------------------
HEAD AT TIME  10  =? 10.62
```

```
S+O#DT/2 = 8651.261

FROM GRAPH DETERMINE HEAD AT
END OF INCREMENT
WHEN READY PRESS RETURN
AND ENTER HEAD WHEN REQUESTED

?
```

```
TIME INCREMENT NO  8
--------------------------
HEAD AT TIME  8  =? 11.61
```

```
S+O#DT/2 = 9922.849

FROM GRAPH DETERMINE HEAD AT
END OF INCREMENT
WHEN READY PRESS RETURN
AND ENTER HEAD WHEN REQUESTED

?
```

```
TIME INCREMENT NO  11
--------------------------
HEAD AT TIME  11  =? 10.0
```

```
S+O#DT/2 = 7937.134

FROM GRAPH DETERMINE HEAD AT
END OF INCREMENT
WHEN READY PRESS RETURN
AND ENTER HEAD WHEN REQUESTED

?
```

```
TIME INCREMENT NO  9
--------------------------
HEAD AT TIME  9  =? 11.20
```

```
S+O#DT/2 = 9358.416

FROM GRAPH DETERMINE HEAD AT
END OF INCREMENT
WHEN READY PRESS RETURN
AND ENTER HEAD WHEN REQUESTED

?
```

```
TIME INCREMENT NO  12
--------------------------
HEAD AT TIME  12  =? 9.3
```

```
S+O#DT/2 = 7064.925

FROM GRAPH DETERMINE HEAD AT
END OF INCREMENT
WHEN READY PRESS RETURN
AND ENTER HEAD WHEN REQUESTED

?
```

TIME INCREMENT NO 13

HEAD AT TIME 13 =? 8.42

TIME INCREMENT NO 16

HEAD AT TIME 16 =? 5.48

S+O*DT/2 = 6029.983

FROM GRAPH DETERMINE HEAD AT
END OF INCREMENT
WHEN READY PRESS RETURN
AND ENTER HEAD WHEN REQUESTED

?

S+O*DT/2 = 4254.092

FROM GRAPH DETERMINE HEAD AT
END OF INCREMENT
WHEN READY PRESS RETURN
AND ENTER HEAD WHEN REQUESTED

?

TIME INCREMENT NO 14

HEAD AT TIME 14 =? 7.4

HEAD IN LAST INCREMENT = ? 5.3

S+O*DT/2 = 5118.321

FROM GRAPH DETERMINE HEAD AT
END OF INCREMENT
WHEN READY PRESS RETURN
AND ENTER HEAD WHEN REQUESTED

?

FINAL RESULTS

TIME	INFLOW	OUTFLOW	HEAD
0	1000	1001.47	5.52
3	1200	1078.63	5.8
6	1500	1330.2	6.67
9	2000	1721.15	7.92
12	2600	2296.87	9.599999
15	2900	2778.88	10.9
18	3100	3046.88	11.59
21	3000	3054.77	11.61
24	2800	2894.39	11.2
27	2600	2672.49	10.62
30	2300	2441.91	10
33	2100	2190.05	9.3
36	1700	1886.68	8.42
39	1400	1554.45	7.4
42	1100	1279.68	6.5
45	1000	990.61	5.48
48	1000	942.2	5.3

TIME INCREMENT NO 15

HEAD AT TIME 15 =? 6.5

S+O*DT/2 = 4480.487

FROM GRAPH DETERMINE HEAD AT
END OF INCREMENT
WHEN READY PRESS RETURN
AND ENTER HEAD WHEN REQUESTED

?

Program notes

(1) Lines 100-120 — clear screen and print title.
(2) Lines 140-280 — enter basic hydrograph data and dimension arrays.
(3) Lines 300-380 — enter hydrograph ordinates.
(4) Lines 390-580 — print entered data and check to ensure it is all correct.
(5) Line 590 — transfers to subroutine for plotting graph.
(6) Lines 600-1020 — calculation loop.
(7) Lines 620-640 — print title.
(8) Lines 660-700 — enter head at beginning of time increment.
(9) Lines 720-790 — transfer to subroutine for storage and outflow; then back.
(10) Line 800 — calculates $O_2\Delta t/2 + S_2$.
(11) Lines 820-860 — print value of $O_2\Delta t/2 + S_2$ and save head and outflow for later print out.
(12) Lines 880-980 — print instructions referring operator to determine H_2 from previously plotted graph (see 590 and subroutine 5000-5420).
(13) Line 1015 — transfers to subroutine for last time increment.
(14) Lines 1040-2000 — print final results and stop execution.
(15) Lines 5000-5420 — subroutine for plotting graph of H and $O\Delta t/2 + S$.
(16) Lines 5000-5200 — print title and instructions.
(17) Line 5220 — enters chosen value of head.
(18) Line 5230 — tests for completion of graph.
(19) Line 5240 — calculates $O\Delta t/2 + S$.
(20) Lines 5280-5310 — print co-ordinates of point on graph.
(21) Line 5320 — creates loop for additional calculations.
(22) Lines 5340-5420 — check if graph is complete.
(23) Lines 5400-5420 — clear screen and return.
(24) Lines 6000-7020 — subroutine for outflow and storage.
(25) Lines 8000-8120 — subroutine to handle last time increment.

Note: Data for the graph of $S + O\Delta t/2$ against H are given in the run. The graph itself is not plotted here. Minor discrepancies in outflow at the end of the time period are due to inaccuracies in reading the graph.

PROBLEMS

(8.1)   A catchment area of 750 hectares is being considered for a water supply to a town of 80 000 people. The mean demand throughout the year may be taken to be 250 l/head/day. The table below gives the available runoff for the dryest period to be considered.

Month	Oct.	Nov.	Dec.	Jan.	Feb.	Mar.	Apr.	May
Run-off	16.5	11.4	10.2	4.0	4.3	4.4	7.6	10.9
(cm over catch)								

Month	June	July	Aug.	Sept.	Oct.	Nov.	Dec.	Jan.
Run-off	11.9	9.1	6.1	5.0	5.1	9.1	11.1	11.7
(cm over catch)								

Find the minimum storage required in order to meet the demand over the period for which information is given.

If site conditions limit the capacity of the storage reservoir to 333 000 m^3 over what period will it be necessary to augment the supply from some other source in order to meet the demand at all times? If no other source of supply is available how many litres/head/day can be supplied?

(8.2) A stream has a minimum dry weather flow of 1.0 m^3/s. It is to be used for electrical power supply, the effective head being 50 m and the overall efficiency 0.8. Calculate the storage necessary to meet the daily demand described below.

Time (h)	0	3	6	9	12	15	18	21	24
Demand (kW)	220	120	250	550	660	560	450	400	220

(8.3) Example 8.2 uses subroutines to calculate outflow and storage according to Equations (8.14) and (8.15). Modify the program to cope with more complex situations where the outflow-elevation and storage-elevation relationships are provided in graphical form. Check your program by running Example 8.2 again but using plotted data instead of Equations (8.14) and (8.15).

(8.4) A reservoir discharges over a spillway 20 m long. It may be assumed that the flow over the spillway crest is critical and that there are no losses. The surface area of the reservoir is 3 km^2 and there is no inflow during the period under consideration. If the upstream water surface is initially 1 m above the spillway crest, determine the level of the water surface after three hours. Use an analytical solution and then write a program to solve the problem numerically with a view to finding a suitable time increment to keep errors to an acceptable level.

Index